NATURAL SELECTION
Methods and Applications

NATURAL SELECTION
Methods and Applications

Mario A. Fares

School of Genetics and Microbiology
Department of Genetics
University of Dublin
Trinity College, Dublin, Ireland

CRC Press
Taylor & Francis Group
Boca Raton London New York

CRC Press is an imprint of the
Taylor & Francis Group, an **informa** business

A SCIENCE PUBLISHERS BOOK

Cover illustration reproduced by courtesy of Dr. Mario A. Fares (Editor) and Dr. Christina Toft.

CRC Press
Taylor & Francis Group
6000 Broken Sound Parkway NW, Suite 300
Boca Raton, FL 33487-2742

First issued in paperback 2019

© 2015 by Taylor & Francis Group, LLC
CRC Press is an imprint of Taylor & Francis Group, an Informa business

No claim to original U.S. Government works

ISBN-13: 978-1-4822-6372-5 (hbk)
ISBN-13: 978-0-367-37813-4 (pbk)

Library of Congress Cataloging-in-Publication Data

Fares, Mario A., author.
    Natural selection : methods and applications / Mario A. Fares.
    p. ; cm.
    Includes bibliographical references and index.
    ISBN 978-1-4822-6372-5 (hardcover : alk. paper)
    I. Title.
    [DNLM: 1. Selection, Genetic. 2. Evolution, Molecular. QU 475]

QH390
572.8'38--dc23   2014026022

Visit the Taylor & Francis Web site at
http://www.taylorandfrancis.com

and the CRC Press Web site at
http://www.crcpress.com

Dedication

To my parents Kamel and Teresa,
to my daughter Lucia

Preface

Since the discovery of the main principles of Natural Selection by Charles Darwin, researchers have been pursuing the identification of the mechanisms that regulate organism's adaptations. It took an extraordinary sense of observation by Darwin and 150 years of research to realize about two of the main principles characterizing natural selection: the slow pace at which new adaptations emerge and that adaptive evolution, that is the emergence and fixation of new traits improving organism's fitness, is subtle and many times impossible to spot. The principles emerging from the study of natural selection have fascinated evolutionary biologists for over a century. The fingerprints of this force of adaptation has been sought by many and the existence of a designing hand behind it has been dismissed by the evolutionary morphological "monstrosities" emerging in nature. Another leap in the understanding of natural selection was made by the use of molecular data emerging from the sequencing of genes and the realization that mutations in protein-coding genes are of stochastic nature and that only those with neutral effects on protein's function survive in the genome. Kimura formalized in a mathematically elegant manner how important neutral mutations are in the generation of genetic variability. Today we know that such genetic variability is a reservoir of possible future adaptations as many variants can be enriched in the population to a point so that when combined in a single genome can lead to novel adaptations.

The neutral theory that rests on the importance of neutral mutations in the generation of genetic variability has led to a better understanding of the complexities of natural selection. Such complexities comprise more than two types of mutations: Neutral mutations, adaptive changes, and those mutations that are neutral but bear potential for novel adaptations (also known as exaptations). This intermingled nature of natural selection has made it difficult to disentangle neutral evolution signatures from those leading to adaptations.

In recent years, a plethora of genomes have been sequenced and new genomes are being released at an unprecedented pace. In addition to this, protein structures are enriching our databases and a bewildering set of network analyses are being performed at many different complexity levels, including protein-protein interactions, metabolic reactions, transcriptomics, etc. Integrating this information to infer the signatures of natural selection requires highly sensitive, fast, and accurate methods and computational tools to deal with the burst of data, together with an absolute understanding of how natural selection works. Unfortunately, finding evidence of adaptive evolution remains a difficult task as methods to deal with data and the models to reproduce the most important parameters governing the action of natural selection are in their infancy. Despite limitations, currently it is imperative to have a complete understanding of the models and methods to read the history of genes and organisms.

Chapters that guide us into the methods designed for the identification of natural selection form this book. It is divided into 9 chapters. The first chapter is a brief introduction to the importance of identifying natural selection and how can signatures of natural selection be found taking into account that distribution of individual's fitness effects in a fitness landscape. Basic concepts on genes' structure, mutations and nucleotide and codon biases are given in the second chapter. These concepts are important for later understanding the ways in which deviations from such structures can help identifying signatures of adaptive evolution. In particular, mutation and nucleotide bias is brought forward as two main parameters in helping to unearth signatures of selection. The basic models of molecular evolution are described in the third chapter. Statistical ways to compare the goodness-of-fit of the models to the data given certain hypothesis are also detailed. Brushstrokes on the neutral theory of molecular evolution are provided in the fourth chapter. In this chapter, we also distinguish two levels of sequence analysis: inter-species and intra-population. The link between natural selection and the origination of novel functions is given in Chapter 5. Indeed, we show that divergence at the amino acid sequence level can lead to functional divergence, often leaving phylogenetic patterned fingerprints. Two methods of functional divergence are discussed in this chapter, one based on a maximum-likelihood framework and another based on a non-parametric approach. Chapter six calls to caution in the application of methods to identify natural selection as methods should account for recombination, which could have important biasing effects in the estimations of the parameters of selection. Once methods to identify natural selection have been examined, chapter seven puts forward the factors that influence the rate of evolution of protein-coding genes. Plausible hypotheses are presented in this chapter to explain the variation of protein's evolutionary rate. Selective pressures in the context of molecular networks are presented in Chapter 8. Finally, how do mutations interact and the consequences of these interactions in the emergence of adaptations at the molecular level are discussed and exemplified in Chapter 9.

This book collects the main principles underlying the action of natural selection and describes what methods are generally used to identify its action. In particular, a useful guide in the identification of the signatures of natural selection is given in this book, targeting students of the molecular biology field interested in evolution.

I would like to thank all the authors that have contributed to this book. I also wish to thank anonymous reviewers as well as the editors who made possible the publication of the book.

Mario A. Fares

Contents

CHAPTER 1

The Role of Natural Selection in Evolution

Mario A. Fares

1.1 Introduction

Life hosts a bewildering variety of living forms each colonizing a particular environment. Most of this variety is dynamic with new forms emerging through the modification of ancestral ones, an observation that constitutes the principle of "Descendent with modification" (Darwin 1859). Biologists have long been enthralled with the idea that the origin of life can be found by linking all life on earth in the form of a phylogenetic tree (Haeckel 1866). These attempts were based upon the principle that a single ancestor for all of life differentiated into multiple descendant species, each adapting by natural selection to one of a new set of ecological niches (see for review (Losos and Schluter 2000; Schluter 2000)). But, how does natural selection act?

School of Genetics & Microbiology, Dept. of Genetics, University of Dublin, Trinity College, Dublin 2, Dublin, Ireland.

In his remarkable book "On the Origin of Species" (Darwin 1859), Charles Darwin emphasized his observations on the variation of natural forms and attributed this to an underlying force "Natural selection". By this consideration, he challenged views followed at that time, that argued that inherited adaptations and coadaptations of species could not be attributed to external environmental conditions, neither could the merit of coevolution between different forms be adduced to forces of creation, claims that he qualified as "preposterous". Ever since his first observations, Charles Darwin realized the need for uncovering the mechanism and means of modification and coadaptation. His observations and beliefs on the existence of such a force were made explicit by showing that variation within species in nature, both for plants and animals, was orders of magnitude lower than in domesticated species. In his explanation of this striking difference, Charles Darwin compared the hand of man to that of natural selection in selecting this variability, but as man selects for new characters, others manifested as byproducts of selection known as "Evolutionary monstrosities".

Despite careful observations by many research groups for many years since the publication of "On the Origin of Species", the role of natural selection on the origin and diversification of species remains controversial, making it a fundamental objective of evolutionary biology. The controversy arises as a result of the complexity of natural selection dynamics, whose defenders struggle to explain the origin of a new species from two populations living in the same context. Although Darwin and others, such as Wallace, Jordan and Wagner, wrestled on the adaptation of populations and their geographical isolation to explain the origin of species, the key observation made by Darwin— that is, little success of some hybrids, descending from the cross of individuals belonging to different populations, to

2

survive—led to a second period of thinking that constituted the Modern Synthesis. The founders of such a period, chief among which were Dobzhansky and Mayr, found that species are characterized by their reproductive isolation from one another. This reproductive isolation can either happen at the prezygotic or postzygotic levels (see Orr and Presgraves 2000). Since the mechanism of natural selection relies on the assumption that individuals unfit to their environment should be removed by natural selection, this force is deemed to oppose itself to the existence or selection of reproductive isolation by the generation of hybrids. Nonetheless, reproductive isolation is a fundamental driver of speciation of populations living in **sympatry**. This fact, how selection could favor the evolution of hybrids sterility, haunted Darwin and his colleagues for years, in particular because selection could not allow for phenotypes to cross fitness valleys but would restrict their existence in adaptive peaks (Figure 1.1). Given this tantalizing fact, understanding the role of selection in evolution requires a detailed description of the relationship between divergence, reproductive isolation and selection.

1.2 Divergence and Evolution by Natural Selection

The importance of selection against hybrids was first appreciated by Wallace (Wallace 1889), who claimed that group selection could in fact cause hybrid sterility and lead to the diversification of species living in sympatry. Does divergence imply evolution? The answer to this question is difficult; the key lies on the relationship between divergence and reproductive isolation. Selection plays an important and direct role in the evolution of sympatry—i.e., the divergence between populations required for them to coexist. However,

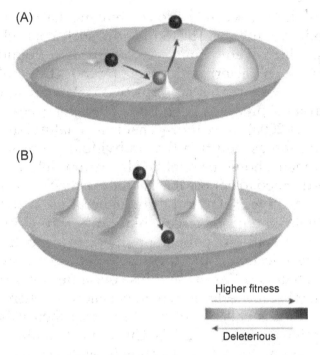

(A)

(B)

Higher fitness

Deleterious

Figure 1.1. Fitness landscape of an evolving population. Peaks represent regions of maximum relative biological fitness while valleys are regions of low fitness. In a smooth landscape (A) populations (spheres) can cross the valleys of low fitness without leading to lethal phenotypes (e.g., these populations go from high mean fitness to intermediate mean fitness). In a rugged and complex landscape (B), crossing the valleys of low fitness is lethal and precludes populations from reaching new local fitness maxima through gradual evolution.

Color image of this figure appears in the color plate section at the end of the book.

the role adduced to reproductive isolation in splitting one species into two (speciation) is dependent upon the biological divergence of the coexisting populations. It is therefore the combination of post-, prezygotic isolations and ecological divergence that limits recombination between populations (gene flow), hence favoring speciation (Barton 2010).

Evolution of speciation by reproductive isolation was first described by the Dobzhansky-Muller model (Dobzhansky 1937; Muller 1940; Muller 1942), according to which this is possible by assuming that postzygotic isolation results from the interaction of two or more genes. That is, Natural selection may have not tested all possible combinations of those alleles that arise by mutation. This model assumes, therefore, that newly arising alleles in two different populations can be perfectly viable and fertile in their own genetic background but sterile or even deleterious when brought together into the same genetic background due to their epistasis (interaction). Under this view, hybrids can emerge and evolve without each population having to pass through fitness valleys, with epistasis being a first rule for the Dobzhansky-Muller model (Orr et al. 1997). The particularities of this rule are defined by the dominance relationships between alleles—also known as the recessicivity of speciation alleles (those alleles of one species that are not viable in the genetic background of another species). According to this rule, a species allele is viable in heterozygosis in another specie's genetic background but not in homo- or hemizygosis (Turelli and Orr 1995). What remains to be discovered, however, is what determines the origination and subsequent fixation of new life forms or traits. Despite decades of intense research and heated debates between those defending the casual origin of traits versus those attributing this merit to the action of selection, we remain far from a clear answer. In this book, authors try to show that the line separating the conceptual foundation of neutralism and selection may be too thin to separate both these apparently opposing forces. Nonetheless, when reaching the end of this book my hope is that readers and researchers of the force of selection realize that neutral evolution can allow the emergence and fixation of cryptic genetic variation upon which natural selection will act and allow the fixation of novel traits. From my perspective, neutralism and selection are the two sides of a coin being continuously flipped, and the speed with which the coin flips, controls the rate of evolution.

1.3 Populations Wandering in a Fitness Landscape

The concept of fitness landscapes is a metaphoric representation of the distribution of fitness effects for different genotypes. There are two main elements in this landscape (Figure 1.1), the peaks refer to genotypes with high reproductive value, and thus high fitness, and the valleys represent regions or genotypes with low fitness. The tendency of genotypes in populations is always to change to those that maximize the fitness of a population, hence crossing fitness valleys in a search for a hill to climb. The merit of the metaphoric representation of genotypes in fitness landscape was of Haldane (Haldane 1931) and Wright (Wright 1932). Wright's diagram represented the different ways in which a genotype could explore phenotypes with differing finesses. In this diagram he envisioned different evolutionary scenarios, all of which highlighted his shifting balance theory of evolution (Wright 1931). Under this theory, genetic drift plays a big role in enabling populations to explore their adaptive landscapes, climbing the different peaks (fitness maxima) and crossing valleys of low fitness. The landscape metaphor has allured evolutionary biologists for decades because it graphically depicts basic evolutionary concepts. One such concept is the speed at which evolution proceeds and the understanding of the counterintuitive destabilizing, even deleterious, nature of mutations that lead to novel traits. Consider for instance a population at the tip of one of many of the adaptive peaks, hence fully adapted to its environment. In such a population, novel mutations, in particular those leading to functional diversification, will push the population downhill towards valleys of low fitness, because these mutations drive individuals away from the ancestral well-adapted functions (Figure 1.1). One way, therefore, to accelerate evolution and emergence of alternative phenotypes is to leap between adaptive peaks, thereby avoiding crossing low-fitness valleys (Figure 1.1). These adaptive leaps are possible through mechanisms of mutational robustness, which ameliorate the effects of mutations, and thus allow smoothing down the abruptness of fitness landscapes (Figure 1.1A compared to Figure 1.1B).

Many studies have illustrated the mechanisms conferring robustness to deleterious mutations and how they allow evolutionary leaps between adaptive peaks. Such mechanisms include molecular chaperones, which fold proteins and their mutated versions in the cell into their functional conformations, gene duplication and genetic interactions. Since such mechanisms ameliorate the deleterious effects of mutations, they confer robustness to mutation effects allowing them to persist in populations and generate exaptations. Robustness therefore is a mechanism that fuels evolution (Draghi et al. 2010). In the following chapters we will explore the consequences of such adaptive accelerations by applying methods to identify signatures of functional divergence. It is important to keep in mind that the purpose of this book, and its contents, is to walk through the different methodologies to identify selection signatures in a pragmatic way, rather than digging into the theoretical constructs of such methodologies. As such, this book should present a good introduction to methods that can be applied to identify Natural selection in protein-coding genes.

7

References

Barton, N.H. 2010. What role does natural selection play in speciation? Philos Trans R Soc Lond B Biol Sci 365: 1825–1840.

Darwin, C. 1859. On the Origin of Species by Means of Natural Selection, or the Preservation of Favoured Races in the Struggle for Life. Murray J., London.

Dobzhansky, T. 1937. Genetics and the origin of species. Columbia, Columbia University Press.

Draghi, J.A., T.L. Parsons, G.P. Wagner and J.B. Plotkin. 2010. Mutational robustness can facilitate adaptation. Nature 463(7279): 353–355.

Haeckel, E. 1866. Generelle Morphologie der Organismen. Berlin, Georg Riemer.

Haldane, J.B. 1931. Mathematical Darwinism: A discussion of the genetical theory of natural selection. Eugen Rev 23: 115–117.

Losos, J.B. and D. Schluter. 2000. Analysis of an evolutionary species-area relationship. Nature 408: 847–850.

Muller, H.J. 1940. Bearing of the Drosophila work on systematics. The new systematics 84.

Muller, H.J. 1942. Isolating mechanisms, evolution and temperature. Biol Symp 55.

Orr, H.A. et al. 1997. The developmental genetics of hybrid inviability: a mitotic defect in Drosophila hybrids. Genetics 145: 1031–1040.

Orr, H.A. and D.C. Presgraves. 2000. Speciation by postzygotic isolation: forces, genes and molecules. Bioessays 22: 1085–1094.

Schluter, D. 2000. The ecology of adaptive radiation. New York, Oxford University Press.

Turelli, M. and H.A. Orr. 1995. The dominance theory of Haldane's rule. Genetics 140: 389–402.

Wallace, A.R. 1889. Darwinism. Mcmillan. London, UK.

Wright, S. 1931. Evolution in Mendelian Populations. Genetics 16: 97–159.

Wright, S. 1932. The roles of mutations, inbreeding, crossbreeding, and selection in evolution. Sixth International Congress on Genetics. pp. 355–366.

CHAPTER 2

Identifying Evolution Signatures in Molecules

Mario A. Fares

2.1 Introduction

The morphological, functional, and some behavioral features of all organisms, including viruses, are codified in their deoxiribonucleic acid (DNA) and ribonucleic acid (RNA). The discovery of this commonality of all life forms has revolutionized science because it offers the opportunity to answer one of the most fundamental questions sought by evolutionary biologists: Where do living organisms come from and how they evolved throughout the history of earth? The coding of the morphological and physiological information though very simple in essence, is useful in providing statistical power, because the entire life information is contained in a combination of four types of nucleotides: adenine (A), thymine (T; although uracil in RNA: U), cytosine (C) and guanine (G). The order in which these four nucleotides are combined determines the functional unites, proteins, of an organism and hence its biological features, which are distinct to a certain extent

School of Genetics & Microbiology, Dept. of Genetics, University of Dublin, Trinity College, Dublin 2, Dublin, Ireland.

from other organisms. What makes this simple code exciting is that the dynamic combination of the four nucleotides across time guards information on all the events that occurred during the evolution of life and its use could unlock the secrets of the history of earth itself.

While different organisms offer distinct complexities in combining and using these four-letter based codes, all of them share the commonality that their functional unites, proteins, are codified by the combination of the four nucleotides in triplets (also known as codons). Each codon forms an amino acid, the molecular building blocks of proteins. This means that, the genetic material could potentially codify for up to 64 different amino acids corresponding to 4^3 combinations of three nucleotides. Conversely, only 20 amino acids form the whole set of known proteins, meaning that some of the triplets are apparently redundant—that is, they codify the same amino acid. As a case in point, the amino acid Glycine is codified by four different codons (Figure 2.1)

		Second Letter										
		U		C		A		G				
1st letter	U	UUU UUC	Phe	UCU UCC	Ser	UAU UAC	Tyr	UGU UGC	Cys	U C	3rd letter	
		UUA UUG	Leu	UCA UCG		UAA UAG	Stop Stop	UGA UGG	Stop Trp	A G		
	C	CUU CUC	Leu	CCU CCC	Pro	CAU CAC	His	CGU CGC	Arg	U C		
		CUA CUG		CCA CCG		CAA CAG	Gln	CGA CGG		A G		
	A	AUU AUC	Ile	ACU ACC	Thr	AAU AAC	Asn	AGU AGC	Ser	U C		
		AUA AUG	Met	ACA ACG		AAA AAG	Lys	AGA AGG	Arg	A G		
	G	GUU GUC	Val	GCU GCC	Ala	GAU GAC	Asp	GGU GGC	Gly	U C		
		GUA GUG		GCA GCG		GAA GAG	Glu	GGA GGG		A G		

Figure 2.1. The degeneracy of the genetic code. While there are 64 possible codons, this only encodes 20 different amino acids, with many codons differing in one or two nucleotides encoding the same amino acid.

Color image of this figure appears in the color plate section at the end of the book.

that differ from one another in the nucleotide of the third codon position. Similar to the letters in the alphabet, the order of many of the amino acids determine the structure and function of proteins. Despite the degeneracy of the genetic code, the combinatorial position of the nucleotides and codons to form the functional and structural units of organisms provides an unlimited source of information. For example, each position of a protein can adopt one of the 20 possible amino acids. The average length of a protein in bacteria is 300 amino acids, hence the possible number of different proteins is 20^{300}, a number that is larger than the number of molecules in the universe. In practice, not all possible combinations are viable and many combinations are constrained by functionality in a network formed by other proteins, environment, and contingency. The number of probable proteins remains, nevertheless, enormous, providing an unlimited source of information.

DNA suffers changes across time owing to errors (mutations) during its replication. These errors are stochastic ("random"), although the fixation of such changes obeys rules that are set by the physico-chemical properties and biological circumstances of the environment and these rules are universal and common to all life on earth. If we were able to determine these rules, and the way in which mutations appear in one group of organisms but not in another, then we would be able to reconstruct the evolutionary history of life through the changes of its molecules. Arguably, mathematical models can be determined to account for these rules and identify the patterns of DNA variation, thereby drawing the way in which a DNA molecule has changed leading to different life forms, ultimately coalescing back to the last common ancestor. Describing the evolution of life is simple, as it is the composition of the code upon which we can base our models, however, there are several complications that limit our capacity to unambiguously relate all the organisms and find the origin of life. First, models that describe the patterns of DNA change are very simplistic because such patterns are ruled by an unlimited, many times unknown, number of environmental factors. Second, there is a large list of different biological phenomena that perturb the rules of DNA evolution upon which models are based, which leads to inaccuracies in the estimates of evolutionary changes and ultimately to a false, although coherent, picture of the history of a set of organisms. For example, models of phylogenetic inference are vulnerable to systematic errors such as long-branch attractions (Felsenstein 1978; Hendy and Penny 1989; Huelsenbeck and Hillis 1993; Huelsenbeck 1995; Philippe 2000; Ruano-Rubio and Fares 2007), gene transfer between unrelated organisms, commonly known as Horizontal Gene Transfer (Lake et al. 1999; Koonin et al. 2001; Wolf et al. 2002), gene duplication and functional divergence within lineages, and other phenomena of which we have little or no knowledge.

Despite the numerous limitations in inferring the evolutionary history of life forms through their molecules, using DNA is advantageous in comparison to classical approaches in which morphological and physiological characters were used to determine the relationship between organisms. First, the evolution of morphology is complex, even for a short lapse of time. Second, in contrast to DNA, which offers an unlimited number of characters or states, the number of morphological and physiological characters is limited. Third, the combinatorial nature of the molecular information allows storing an infinite amount of phylogenetic information, making it ideal for the determination of history of life, a virtually impossible task when using morphological approaches.

The advantage of using molecular information (DNA, RNA and proteins) over the morphological characters makes it feasible to infer fundamental biological and evolutionary processes. In this sense, the field of evolutionary biology has itself evolved, devoting decades of effort to trying, conceiving, and developing models that account for the changes in DNA and proteins and put these models at test by applying them to organisms with known evolutionary history. As will be shown later, these models have revolutionized our view of the evolutionary history of organisms, have broadened research horizons, and opened new avenues in biotechnology and biomedical research.

2.2 Structure of the Genetic Material

Before we start with the exposure of the different models and evolutionary events that DNA and RNA molecules has been through, we need to highlight a few basic concepts that are important for the understanding of these events. The way in which the genetic material is organized is paramount to this understanding because this organization bears an intrinsic complexity that complicate models of evolution. For simplicity, I will concentrate on DNA based organisms (DNA-based viruses, bacteria, animals and plants). By no means, however, do I purposely dismiss RNA based viruses, whose complexity deserves the devotion of an entire book. In general terms, DNA is organized into three classes according to its final product: non-coding DNA, RNA-coding DNA and protein-coding DNA. Non-coding DNA is non-transcribed DNA and its role can be structural or regulatory. The importance of this DNA from the evolutionary perspective remains the ground for intense investigation and several models have been proposed to explain the C-value enigma (see for review

(Gregory 2001). The amount of coding DNA varies sharply among organisms: viruses lack of non-coding DNA, the fraction of non-coding DNA in bacteria is very low so that bacterial genomes are almost exclusively coding, and a large proportion of DNA in eukaryotic organisms is non-coding.

RNA-coding genes do not yield proteins as a final product but rather they produce transfer RNA (tRNA), ribosomal RNAs (rRNA), and small nuclear RNAs (snRNA). Protein-coding genes are those genes that are transcribed into messenger RNA (mRNA), which are subsequently decoded by tRNAs and translated by rRNAs at the core of ribosomes into amino acids, the building blocks of proteins.

Protein-coding genes present different structures according to the organism and this structure is responsible for the complexity underlying the flow of genetic information from genes to proteins. For example, in DNA viruses and bacteria, the information to form proteins is contained within a DNA fragment that is entirely transcribed and translated. In eukaryotes, gene organization is more complex because it intermingles coding units (exons) and non-coding units (introns), the latter of which need to be excised before forming proteins. Explaining the processing of eukaryotic genes is beyond the scope of this book and I will only concentrate on giving a broad view of the flow of genetic information from DNA to proteins.

The nucleotide sequence of a gene that carries the genetic information in a codified form, is transcribed in a nucleotide by nucleotide bases to messenger RNA in prokaryotes, or pre-messenger RNA in eukaryotes, and it is then translated in codon by codon bases into long stretches of amino acids. The translation of each of the codons into an amino acid follows a degenerate genetic code, according to which, with the exception of tryptophan and Methionine, there is a correspondence of many codons to one amino acid.

With only few exceptions, such as vertebrate and non-vertebrate mitochondrial genes, the genetic code is universal. Other exceptions have been reported for bacteria with reduced genome sizes, such as mycoplasmas and endosymbiotic bacteria, in which UGA stop codon has been reassigned to tryptophan (McCutcheon et al. 2009). In the universal genetic code, 61 codons codify for 20 amino acids, while the remaining three codons (UAA, UAG and UGA) pinpoint the termination of translation. The initiation amino acid methionine marks the start of translation, once the polypeptide generated methionine is modified and subsequently removed. Previous studies have shown that CUG (the codon codifying for Leu) and UUG (the codon codifying for Phe) can also be used as initiation codons in some nuclear genes (Elzanowski and Ostell 1996). The degenerate nature of the genetic code is such that nucleotide changes (mutations) at third codon positions warrant preserving the amino acid while this is less so in first positions of all the codons. Changes at second positions of the codons leads to amino acid replacements in the protein. Methionine and tryptophan are codified by zero-fold degenerate codons, hence nucleotide substitutions at any of the three codon positions would involve an amino acid replacement. Codons that codify for the same amino acid are called **synonymous codons,** and nucleotide substitutions that change the codon but not the amino acid are called **synonymous substitutions.** Codons that codify for different amino acids are called **non-synonymous codons** and nucleotide substitutions that change the codon and the amino acid are named **non-synonymous substitutions** (Figure 2.2).

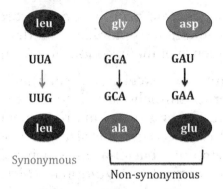

Figure 2.2. Synonymous and non-synonymous nucleotide substitutions. The degeneracy of the genetic code implies that some nucleotide substitutions will lead to codon changes but not to amino acid replacements (synonymous substitutions). Non-synonymous nucleotide substitutions refer to those that change the codon and the encoded amino acid.

Color image of this figure appears in the color plate section at the end of the book.

2.3 Mutations of DNA Sequences

Mutations here are defined as random changes in the nucleotide sequences of genes. Regardless of their origin, spontaneous versus induced, mutations follow a stochastic process and when they affect protein-coding genes they can change the protein amino acid sequence and alter the morphological or physiological characters of organisms. There are many different types of mutations according to the nature and type of change: (a) **punctual** nucleotide substitutions: these are replacements of one nucleotide by one of the remaining three nucleotides; (b) **deletion** of nucleotides: this involves the deletion of one or more nucleotides from the sequence; (c) **insertion** of nucleotides: this involves the insertion of one or more nucleotides to the sequence and (d) **inversion**: which involves the flip

16

over of a segment of the DNA sequence. When mutations affect the reading frame of the nucleotide sequence, in protein-coding DNA sequences for instance, these are called **frame-shiftmutations**. Owing to their effect in shifting the frame of protein-coding genes, Insertions and Deletions of nucleotides (**InDels**) have generally deleterious effects because they disrupt the coding sense generating disrupted protein sequences. Studies on the selective pressures on InDels have been underwhelming, in particular with regards to modeling the evolution of InDels rates. In this book we will focus on modeling DNA punctual mutations, which have been more extensively investigated.

There are many different ways to classify punctual DNA mutations, which are mostly based on the strength of mutations (chemical strength) or the effect that such a mutation has on the protein. For example, from the chemical perspective, nucleotide mutations can be named transitions when they involve the transition between nucleotides that are chemically equivalent (from a purine to a purine, nucleotides A and G, or from a pyrimidine to a pyrimidine, C and T), or transversions, when they involve the replacement between chemically different amino acids (from a purine to a pyrimidine and vice versa) (Figure 2.3). It is known that transitions are more frequent than transversions because the former does not induce dramatic effects on the DNA sequence (Fitch 1967; Gojobori et al. 1982; Kocher and Wilson 1991). Importantly, transitions in the third positions of two-fold degenerate codons lead to synonymous codon replacements while transversions at these positions lead to non-synonymous codon replacements. As for the codon and amino acid substitutions concerns, nucleotide mutations can be classified into synonymous substitutions, when such a substitution involves the change of a codon by another that codifies for the same amino acid, and non-synonymous

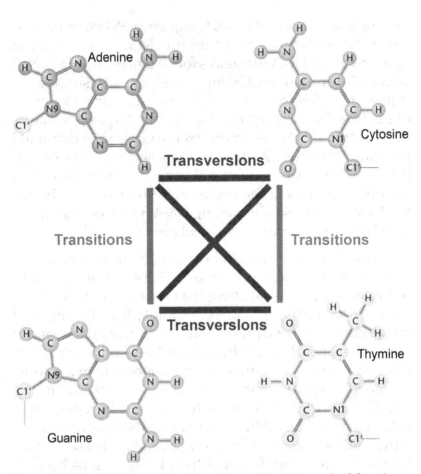

Figure 2.3. Transitions and transversions. Transitions (red lines) are nucleotide substitutions between two chemically similar bases (e.g., Adenosine A to Guanine G or Cytosine C to Thymine T). Transversions (blue lines) are transitions between chemically different nucleotides (e.g., A or G to C or T). While transversions are more numerous, hence likely, these are less favored or more constrained by natural selection as they lead to important changes in the codon composition and, very probably, amino acid composition.

Color image of this figure appears in the color plate section at the end of the book.

substitutions, when the nucleotide replacement involves a change from one codon to another that codifies for a different amino acid. Notice that roughly the proportion of nucleotide substitutions in the genetic code that are non-synonymous (for example, second codon positions and most changes at first codon positions) is larger (71%) than that of synonymous substitutions (25%) and nonsense mutations (those leading to stop codons, 4%) (Nei 1975; Li 1997). In practice, although all substitutions are equally likely owing to the stochastic nature of mutations, not all substitutions lead to viable or competitive **phenotypes**, hence are purified out by Natural selection (see next chapters). For example, non-synonymous nucleotide substitutions lead to variation in the amino acid sequence of proteins and, since mutations are random and devoid of design, they are likely to lead to less-optimal or deleterious phenotypes (for example, badly performing morphologies or physiologies), which will be purged by Natural selection.

2.4 Nucleotide and Codon Biases

In theory, one would expect that proteins of an organism be represented by equal frequencies of each of the codons. For example, the frequency of each of the synonymous codons for a particular amino acid should be the same provided that there is no selection pressure biasing codon representation: for Gly, the probability of each of the codon should be represented by a 25%. In practice, however, the frequency of codons varies dramatically from one organism to another and between proteins (Ikemura 1985). As we will see later in this chapter, there are several factors that account for the difference in the codon usage between organisms, including transcriptional and translational efficiency, translational robustness, and biased mutation pressure. In addition to these factors, codon usage may vary within organisms, in particular bacteria, due to horizontal gene transfer.

With regard to translational efficiency, Ikemura showed that in the bacterium *E. coli* and yeast, codon usage is correlated with translational efficiency: the frequency of codon frequency in highly expressed genes is correlated with the relative abundance of isoaccepting tRNAs in the cell (Ikemura 1981; Ikemura 1985). That is, there is a correspondence between the abundance of tRNAs and that of the codons in highly expressed genes. This correspondence is mainly mediated by the fact that highly abundant tRNAs are codified by genes with high-copy numbers in the cell. Therefore, for highly expressed genes, codon bias should be similar within the same cell or organism but very different between organisms, even between those organisms that are closely related. For example, the heat-shock protein GroEL of *E. coli* presents a very different codon usage table in comparison with other highly related bacteria. In sharp contrast to highly expressed genes, moderately to lowly expressed genes show more equilibrated distributions of synonymous codons because the translation of these genes is slow and rare tRNAs can be used.

Codon usage differs among genes according to their transcriptional and translational efficiency (Grantham et al. 1981; Ikemura 1981; Gouy and Gautier 1982; Grosjean and Fiers 1982; Ikemura 1985). In addition to this, genes horizontally transferred from bacteria with different genome codon usage may disrupt the pattern of codon usage in highly expressed genes (Medigue et al. 1991). These genes preferentially bear the codon usage of their original source, hence varying in their use of codons from the typical and highly expressed genes in the host bacterium. Arguably, lack of correspondence between gene expression and codon usage has been previously used as a signature of horizontal gene transfer (Lawrence and Ochman 1997; Karlin and Brocchieri 2000; Karlin and Mrazek 2000). Also, using the modal of codon usage distribution has proven effective in identifying the source of genes that have been horizontally transferred (Davis and Olsen 2010).

The above observations imply that, in highly expressed genes, codons that do not correspond with abundant tRNAs are eliminated by purifying selection because they are not efficiently translated. Moreover, it has been shown that translational robustness—that is increase of the tolerance of proteins folding to translation errors (see subsequent chapters on the rate of protein evolution), is an important selection constraint to protein evolution and is responsible for highly expressed genes being more conserved throughout different scales of evolution (Drummond et al. 2005; Wilke et al. 2005; Bloom et al. 2006; Wilke and Drummond 2006; Drummond and Wilke 2008; Drummond 2009; Drummond and Wilke 2009; Toft and Fares 2009).

In addition to the abundance of tRNAs as a factor to explain the codon bias observed in many genomes, biased mutational pressures could also lead to such patterns of codon usages. In bacteria, nucleotide usage—that is, the genome percentage of GC (%GC)—varies drastically among different bacteria, ranging between 16%, in species such as *Carsonella ruddii* (Nakabachi et al. 2006), to *Anaeromyxobacter dehalogens*, with a GC-content of 75% (Belozersky and Spirin 1958; Sueoka 1962; Li 1991; Osawa 1995). This substantial variation in genome GC content has been linked to extrinsic environmental factors (Foerstner et al. 2005), such as aerobiosis and temperature (Kagawa et al. 1984; Naya et al. 2002; Zavala et al. 2002; Musto et al. 2004), ultraviolet radiation (Gause et al. 1967; Singer and Ames 1970) and nitrogen fixation (McEwan et al. 1998); and intrinsic factors, including metabolic factors (Martin 1995), DNA coding-sequence length (Oliver and Marin 1996; Xia et al. 2003), genome size (Musto et al. 2006), and whether the bacterium is free-living or not (Rocha and Danchin 2002; Woolfit and Bromham 2003). Despite the fact that many intrinsic factors have been considered to be main players in the GC-content variation, this variation has been largely attributed to the difference in the forward and backward mutation rates of the AT and the GC pairs in nucleotide sequences (Sueoka 1962). Therefore, whether the other factors mentioned above directly influence genome GC-content or they represent the byproduct of mutational bias, remains controversial. Interestingly, even codon sites that are expected to be under weak selection, such as synonymous nucleotide positions in lowly expressed genes, seem to present larger %GC than expected from mutation equilibrium (Deschavanne and Filipski 1995). This is surprising because there is an excess of GC -> AT mutation bias (Deschavanne and Filipski 1995), particularly in bacteria, which is less acute when examining polymorphism at the population level. This is consistent with selection against mutation bias. Moreover, such GC -> AT bias is also observed among pseudogenes of *Mycobacterium leprae* (Balbi et al. 2009). This accumulating evidence supports the role of selection to increase GC-content in the genome (Lynch 2007). Despite the controversy associated with the ultimate reasons behind the excess in GC-rich bacterial species, this seems to be adduced to a selection in favor of increased genome GC-content (Hildebrand et al. 2010). Nevertheless, a recent study supports that DNA polymerase III α subunit and its isoforms that participate in either replication or SOS mutagenesis play a dominant role in determining GC-content variation (Wu et al. 2012). In spite of the ambiguity surrounding the factors governing genome GC-content, it seems clear that different forces drive variation in GC content. For example, GC-content of third codon positions, in principle devoid of strong selection, correlate strongly with genome GC-content, supporting the role of mutational bias in these nucleotide third positions, in which most nucleotide substitutions are synonymous. Conversely, second and first codon positions are more subjected to purifying selection because nucleotide substitutions at these positions are more likely to lead to amino acid replacements. At these positions, GC–content obeys functional constraints rather than mutational biases. Importantly, the patterns of nucleotide substitutions and codon usage bias vary greatly among high-eukaryotic organisms, even at short evolutionary distances correlating with gene length (Moriyama and Powell 1998) and expression (Hiraoka et al. 2009). For example, the invertebrate Drosophila presents quite a strong codon usage bias mostly due to translational efficiency (Shields et al. 1988; Akashi 1994; Moriyama and Powell 1997). In vertebrates, the differential gene expression levels among tissues complicate the understanding of codon bias, with these genomes showing a mosaic distribution of GC-rich and GC-poor regions (known as **isochors**) (Bernardi et al. 1985; Bernardi et al. 1988). The origin of these isochors remains the subject of controversy (see for more detail (Wolfe et al. 1989; Holmquist and Filipski 1994; Bernardi 1995)). An interesting corollary of these studies is that there is dependence between the position of genes in the different isochors and their rates of evolution because codon usage biases affect the rate of nucleotide substitution.

References

Akashi, H. 1994. Synonymous codon usage in Drosophila melanogaster: natural selection and translational accuracy. Genetics 136: 927–935.

Balbi, K.J., E.P. Rocha and E.J. Feil. 2009. The temporal dynamics of slightly deleterious mutations in *Escherichia coli* and *Shigella* spp. Mol Biol Evol 26: 345–355.

Belozersky, A.N. and A.S. Spirin. 1958. A correlation between the compositions of deoxyribonucleic and ribonucleic acids. Nature 182: 111–112.

Bernardi, G. 1995. The human genome: organization and evolutionary history. Annu Rev Genet 29: 445–476.

Bernardi, G., D. Mouchiroud and C. Gautier. 1988. Compositional patterns in vertebrate genomes: conservation and change in evolution. J Mol Evol 28: 7–18.

Bernardi, G., B. Olofsson, J. Filipski, M. Zerial, J. Salinas, G. Cuny, M. Meunier-Rotival and F. Rodier. 1985. The mosaic genome of warm-blooded vertebrates. Science 228: 953–958.

Bloom, J.D., D.A. Drummond, F.H. Arnold and C.O. Wilke. 2006. Structural determinants of the rate of protein evolution in yeast. Mol Biol Evol 23: 1751–1761.

Davis, J.J. and G.J. Olsen. 2010. Modal codon usage: assessing the typical codon usage of a genome. Mol Biol Evol 27: 800–810.

Deschavanne, P. and J. Filipski. 1995. Correlation of GC content with replication timing and repair mechanisms in weakly expressed *E. coli* genes. Nucleic Acids Res 23: 1350–1353.

Drummond, D.A. 2009. Protein evolution: innovative chaps. Curr Biol 19: R740–742.

Drummond, D.A., J.D. Bloom, C. Adami, C.O. Wilke and F.H. Arnold. 2005. Why highly expressed proteins evolve slowly. Proc Natl Acad Sci U S A 102: 14338–14343.

Drummond, D.A. and C.O. Wilke. 2008. Mistranslation-induced protein misfolding as a dominant constraint on coding-sequence evolution. Cell 134: 341–352.

Drummond, D.A. and C.O. Wilke. 2009. The evolutionary consequences of erroneous protein synthesis. Nat Rev Genet 10: 715–724.

Elzanowski, A. and J. Ostell. 1996. The genetic codes. National Centre for Biotechnology Information (NCBI), Bethesda, MD.

Felsenstein, J. 1978. Cases in which parsimony or compatibility methods will be positively misleading. Systematic Zoology 27: 10.

Fitch, W.M. 1967. Evidence suggesting a non-random character to nucleotide replacements in naturally occurring mutations. J Mol Biol 26: 9.

Foerstner, K.U., C. von Mering, S.D. Hooper and P. Bork. 2005. Environments shape the nucleotide composition of genomes. EMBO Rep 6: 1208–1213.

Gause, G.F., Y.V. Dudnik, A.V. Laiko and E.M. Netyksa. 1967. Induction of mutants with altered DNA composition: effect of ultraviolet on Bacterium paracoli 5099. Science 157: 1196–1197.

Gojobori, T., W.H. Li and D. Graur. 1982. Patterns of nucleotide substitution in pseudogenes and functional genes. J Mol Evol 18: 360–369.

Gouy, M. and C. Gautier. 1982. Codon usage in bacteria: correlation with gene expressivity. Nucleic Acids Res 10: 7055–7074.

Grantham, R., C. Gautier, M. Gouy, M. Jacobzone and R. Mercier. 1981. Codon catalog usage is a genome strategy modulated for gene expressivity. Nucleic Acids Res 9: r43–74.

Gregory, T.R. 2001. Coincidence, coevolution, or causation? DNA content, cell size, and the C-value enigma. Biol Rev Camb Philos Soc 76: 65–101.

Grosjean, H. and W. Fiers. 1982. Preferential codon usage in prokaryotic genes: the optimal codon-anticodon interaction energy and the selective codon usage in efficiently expressed genes. Gene 18: 199–209.

Hendy, M.D. and D. Penny. 1989. A framework for the quantitative study of evolutionary trees. Systematic Zoology 38: 13.

Hildebrand, F., A. Meyer and A. Eyre-Walker. 2010. Evidence of selection upon genomic GC-content in bacteria. PLoS Genet 6.

Hiraoka, Y., K. Kawamata, T. Haraguchi and Y. Chikashige. 2009. Codon usage bias is correlated with gene expression levels in the fission yeast Schizosaccharomyces pombe. Genes Cells 14: 499–509.

Holmquist, G.P. and J. Filipski. 1994. Organization of mutations along the genome: a prime determinant of genome evolution. Trends Ecol Evol 9: 65–69.

Huelsenbeck, J.P. 1995. The robustness of two phylogenetic methods: four-taxon simulations reveal a slight superiority of maximum likelihood over neighbor joining. Mol Biol Evol 12: 843–849.

Huelsenbeck, J.P. and D.M. Hillis. 1993. Success of phylogenetic methods in the four-taxon case. Systematic Biology 42: 18.

Ikemura, T. 1981. Correlation between the abundance of *Escherichia coli* transfer RNAs and the occurrence of the respective codons in its protein genes: a proposal for a synonymous codon choice that is optimal for the *E. coli* translational system. J Mol Biol 151: 389–409.

Ikemura, T. 1985. Codon usage and tRNA content in unicellular and multicellular organisms. Mol Biol Evol 2: 13–34.

Kagawa, Y., H. Nojima, N. Nukiwa, M. Ishizuka, T. Nakajima, T. Yasuhara, T. Tanaka and T. Oshima. 1984. High guanine plus cytosine content in the third letter of codons of an extreme thermophile. DNA sequence of the isopropylmalate dehydrogenase of Thermus thermophilus. J Biol Chem 259: 2956–2960.

Karlin, S. and L. Brocchieri. 2000. Heat shock protein 60 sequence comparisons: duplications, lateral transfer, and mitochondrial evolution. Proc Natl Acad Sci U S A 97: 11348–11353.

Karlin, S. and J. Mrazek. 2000. Predicted highly expressed genes of diverse prokaryotic genomes. J Bacteriol 182: 5238–5250.

Kocher, T.D. and A.C. Wilson. 1991. Sequence Evolution of mitochondrial DNA in humans and chimpanzees: Control region and a protein-coding region. New York: Springer-Verlag.

Koonin, E.V., K.S. Makarova and L. Aravind. 2001. Horizontal gene transfer in prokaryotes: quantification and classification. Annu Rev Microbiol 55: 709–742.

Lake, J.A., R. Jain and M.C. Rivera. 1999. Mix and match in the tree of life. Science 283: 2027–2028.

Lawrence, J.G. and H. Ochman. 1997. Amelioration of bacterial genomes: rates of change and exchange. J Mol Evol 44: 383–397.

Li, W.H. 1991. Fundamentals of Molecular Evolution: Sunderland MA.

Li, W.H. 1997. Molecular Evolution. Sunderland, MA.

Lynch, M. 2007. The origins of genome architecture.

Martin, A.P. 1995. Metabolic rate and directional nucleotide substitution in animal mitochondrial DNA. Mol Biol Evol 12: 1124–1131.

McCutcheon, J.P., B.R. McDonald and N.A. Moran. 2009. Origin of an alternative genetic code in the extremely small and GC-rich genome of a bacterial symbiont. PLoS Genet 5: e1000565.

McEwan, C.E., D. Gatherer and N.R. McEwan. 1998. Nitrogen-fixing aerobic bacteria have higher genomic GC content than non-fixing species within the same genus. Hereditas 128: 173–178.

Medigue, C., T. Rouxel, P. Vigier, A. Henaut and A. Danchin. 1991. Evidence for horizontal gene transfer in *Escherichia coli* speciation. J Mol Biol 222: 851–856.

Moriyama, E.N. and J.R. Powell. 1997. Codon usage bias and tRNA abundance in Drosophila. J Mol Evol 45: 514–523.

Moriyama, E.N. and J.R. Powell. 1998. Gene length and codon usage bias in Drosophila melanogaster, Saccharomyces cerevisiae and *Escherichia coli*. Nucleic Acids Res 26: 3188–3193.

Musto, H., H. Naya, A. Zavala, H. Romero, F. Alvarez-Valina and G. Bernardi. 2004. Correlations between genomic GC levels and optimal growth temperatures in prokaryotes. FEBS Lett 573: 73–77.

Musto, H., H. Naya, A. Zavala, H. Romero, F. Alvarez-Valin and G. Bernardi. 2006. Genomic GC level, optimal growth temperature, and genome size in prokaryotes. Biochem Biophys Res Commun 347: 1–3.

Nakabachi, A., A. Yamashita, H. Toh, H. Ishikawa, H.E. Dunbar, N.A. Moran and M. Hattori. 2006. The 160-kilobase genome of the bacterial endosymbiont Carsonella. Science 314: 267.

Naya, H., H. Romero, A. Zavala, B. Alvarez and H. Musto. 2002. Aerobiosis increases the genomic guanine plus cytosine content (GC%) in prokaryotes. J Mol Evol 55: 260–264.

Nei, M. 1975. Molecular population genetics and evolution. North Holland, Amsterdam. The Netherlands.

Oliver, J.L. and A. Marin. 1996. A relationship between GC content and coding-sequence length. J Mol Evol 43: 216–223.

Osawa, S. 1995. Evolution of the Genetic Code. Oxford, U.K.: Oxford University Press.

Philippe, H. 2000. Opinion: long branch attraction and protist phylogeny. Protist 151: 307–316.

Rocha, E.P. and A. Danchin. 2002. Base composition bias might result from competition for metabolic resources. Trends Genet 18: 291–294.

Ruano-Rubio, V. and M.A. Fares. 2007. Artifactual phylogenies caused by correlated distribution of substitution rates among sites and lineages: the good, the bad, and the ugly. Syst Biol 56: 68–82.

Shields, D.C., P.M. Sharp, D.G. Higgins and F. Wright. 1988. "Silent" sites in Drosophila genes are not neutral: evidence of selection among synonymous codons. Mol Biol Evol 5: 704–716.

Singer, C.E. and B.N. Ames. 1970. Sunlight ultraviolet and bacterial DNA base ratios. Science 170: 822–825.

Sueoka, N. 1962. On the genetic basis of variation and heterogeneity of DNA base composition. Proc Natl Acad Sci U S A 48: 582–592.

Toft, C. and M.A. Fares. 2009. Selection for translational robustness in Buchnera aphidicola, endosymbiotic bacteria of aphids. Mol Biol Evol 26: 743–751.

Wilke, C.O., J.D. Bloom, D.A. Drummond and A. Raval. 2005. Predicting the tolerance of proteins to random amino acid substitution. Biophys J 89: 3714–3720.

Wilke, C.O. and D.A. Drummond. 2006. Population genetics of translational robustness. Genetics 173: 473–481.

Wolf, Y.I., I.B. Rogozin, N.V. Grishin and E.V. Koonin. 2002. Genome trees and the tree of life. Trends Genet 18: 472–479.

Wolfe, K.H., P.M. Sharp and W.H. Li. 1989. Mutation rates differ among regions of the mammalian genome. Nature 337: 283–285.

Woolfit, M. and L. Bromham. 2003. Increased rates of sequence evolution in endosymbiotic bacteria and fungi with small effective population sizes. Mol Biol Evol 20: 1545–1555.

Wu, H., Z. Zhang, S. Hu and J. Yu. 2012. On the molecular mechanism of GC content variation among eubacterial genomes. Biol Direct 7: 2.

Xia, X., Z. Xie and W.H. Li. 2003. Effects of GC content and mutational pressure on the lengths of exons and coding sequences. J Mol Evol 56: 362–370.

Zavala, A., H. Naya, H. Romero and H. Musto. 2002. Trends in codon and amino acid usage in Thermotoga maritima. J Mol Evol 54: 563–568.

CHAPTER 3

Modeling Evolution of Molecular Sequences

Mario A. Fares

3.1 Introduction

Evolution of life on earth has been inferred using many different sources, all of which have offered an angled view of evolution by natural selection. Classical theories on the evolution of species diversification have relied on observations made from fossil records. Due to its fragmentary nature, fossil records have offered little brushstrokes on the main events of the evolutionary history of organisms. Later, morphological and physiological studies opened new avenues in the understanding of the evolutionary relationships between organisms. However, this data offers a very complex evolutionary picture of organisms, which hampers a detailed identification of the origin of biological novelties, with many theories stemming from this data being brought into the arena of controversy.

Advances in the technology of molecular biology in recent decades have changed this situation. DNA

School of Genetics & Microbiology, Dept. of Genetics, University of Dublin, Trinity College, Dublin 2, Dublin, Ireland.

sequencing, comparative genetics and genomics, fueled by the astonishing growth of genome sequencing technology, and network biology, have astronomically increased our power to test evolutionary theories and to generate new ones. In particular, this kind of data has generated new fields in evolutionary biology, all of which have been aimed at understanding how novel forms of life emerged and the evolutionary dynamics of the interface between organisms and their surrounding environment. The understanding of the history of life viewed through molecular data relies on the assumption that natural selection acts on organisms depending on their performance, which in itself depends on the integrative work of its molecular components codified by the molecule storing the blueprint of all organisms (DNA or RNA in some viruses). Since the code that makes up life forms is unique and common to all organisms (namely, it is formed by four types of nucleotides: adenine (A), guanine (G), thymine (T) and cytosine (C)), using DNA allows comparing organisms with unrelated physiologies or morphologies, such as prokaryotes and multi-cellular eukaryotes. The drawback of using molecular data, however, is that the evolutionary processes underlying such molecular changes are complex and many times difficult to disentangle from stochastic variation. Accordingly, a number of statistical methods have been devised in recent years to identify signatures of natural selection, all of which attempt to apply realistic assumptions and parameterize the evolutionary history of molecules.

Modeling nucleotide evolution, or the substitutions of the nucleotides of a DNA sequence by other nucleotides throughout evolution, is important both conceptually and methodologically. First, measuring nucleotide substitutions provides an idea of how complex and what consequences the mutation of a particular nucleotide site has on: (a) the regulatory expression of a gene (if the mutation happens to occur in the promoter of such a gene), (b) on the structure of a protein (if the mutation affects a protein-coding DNA sequence and yields a different amino acid sequence), and (c) on the tempo and mode of evolution of a particular species. Second, measuring nucleotide differences between species is the first step in building distance matrices that are crucial to draw and understand the phylogenetic relationships between species—ultimately determining the natural history of a collection of species.

As we have seen in the previous chapter, there are many factors that can complicate the description of the mutational patterns of nucleotides. First, the transitions between the four nucleotides, although they can occur at equal rates, are not equally acceptable owing to chemical and mutational constraints. Second, the rate of DNA evolution (substitutions per nucleotide position) is not the same across the genome, with this rate varying among isochors, coding and non-coding DNA, and between codon positions within protein coding genes. As we will see in subsequent chapters, the rates of evolution can also vary among codons for the same position, even for second positions, which are non-degenerate. This variation is much more complex to account for and it is subjected to higher-level constraints, such as functional and structural constraints on the protein variant resulting from these nucleotide substitutions. Modeling nucleotide substitutions is therefore complex, with most models often yielding simplistic estimates of the evolutionary distance between the **homologous** nucleotide sequences of two species descending from the same common ancestor. Taking these limitations into account, I will devote the rest of the chapter to describe models of DNA evolution and practical approaches to identify the best fitting model to explain the divergence of two or more sequences from a last common ancestor.

3.2 Measuring the distance between two nucleotide sequences

Essentially, the distance between two nucleotide sequences can be measured as the number of nucleotide substitutions accounted between both of the sequences (i and j) since their divergence from a common ancestor. This distance can be simplistically estimated as the absolute number of nucleotide substitutions (nd_{ij})

$$d_{ij} = nd_{ij} \qquad (3.1)$$

However, on an average the number of differences varies, according to the length of the sequences—that is the longer the sequence the larger are the number of target sites for mutations. It is desirable to estimate the proportion of differences between two sequences as:

$$\hat{p} = nd_{ij}/n \qquad (3.2)$$

Notice that this proportion ($\hat{p}$) is an estimate of the real distance because by using it we are assuming that all nucleotide sites are equally probable to undergo mutation—that is the likelihood of observing a nucleotide mutation increases as we move further across the sequence. This distance, called the P-distance, has a variance associated to $\hat{p}$ estimate that follows a binomial distribution:

$$V(\hat{p}) = p(1-p)/n \qquad (3.3)$$

While this simple model can provide good estimates of the divergence between two nucleotide sequences, the model becomes simplistic when the sequences compared are too divergent or they present complex nucleotide mutation dynamics.

3.2.1 Models of DNA Evolution: Estimating the distance between nucleotide sequences

When the distance between nucleotide sequences is large, then $\hat{p}$ may not be a good estimate of the real number of nucleotide substitutions between two sequences as this distance does not account for parallel substitutions between phylogenetically unrelated sequences or for reversed substitutions—for example, substitution at time (t) of a nucleotide to its previous nucleotide state at time (t-1). In the case of nucleotide sequences, this problem becomes serious when the divergence between sequences commenced a long evolutionary time ago because there are only four possible nucleotides for a site and the probability of reversible substitution is large. This problem is more dramatic if we take into account that the probability of substitutions between the four nucleotides is not equal. For example, transitions (substitutions between purines or pyrimidines) are more likely than transversions (as mentioned in Chapter 2, page 17). Also, nucleotide substitutions at different positions of a gene or genome region may be differently accepted by natural selection. It is also known that substitutions are variable between the codons of a protein-coding sequence (Gu and Li 1998) according to codon compositional bias and selection constraints. Other factors that are unknown may also contribute significantly to variation in the probability of a substitution of one nucleotide by another. Because of this, a lot of effort has been devoted to develop models that attempt to draw the mutational history of nucleotide substitutions based upon known influencing factors, and to make these models robust to the violation of the assumptions owing to uncontrolled factors. Below, I will briefly describe some of the most used models of nucleotide substitutions and how these models are used in nucleotide sequences in general and in protein-coding sequences in particular. Many more models besides the ones I will describe hereafter have been developed. For detailed description of these models I would refer readers to several other studies and textbooks (Zharkikh 1994; Swofford et al. 1996). Formal mathematical details of the models I will describe herein are also treated in detail elsewhere (Yang 2006).

3.2.1.1 The method of Jukes and Cantor

The JC method (Jukes and Cantor 1969) is the simplest of the models and assumes that at any nucleotide position the transition from one nucleotide occurs with the same probability to any of the three other nucleotides. Because this is the basic model upon which other models are built, I will devote more detail to the mathematical development of this model and will refer, when explaining other models, to the differences that such models present in comparison to JC. Therefore, assuming a probability transition between nucleotides i and j per unit time of α, with i and j being A, T, C or G, one could assume that the probability of the transition of a nucleotide to any of the other three per unit time is equivalent to the rate of substitution per site per unit time and is

$$r = 3\alpha \qquad (3.4)$$

Conversely, the probability of a nucleotide conserving its state after a unit time has elapsed is:

$$q_{ii} = 1 - 3\alpha \qquad (3.5)$$

This is equivalent to $q_{ii} = 1 - r$.

At any single time therefore, the resulting substitution matrix is:

$$Q = \{q_{ij}\} = \begin{bmatrix} 1 - 3\alpha & \alpha & \alpha & \alpha \\ \alpha & 1 - 3\alpha & \alpha & \alpha \\ \alpha & \alpha & 1 - 3\alpha & \alpha \end{bmatrix} \tag{3.6}$$

This matrix bears several interesting features that are of importance. Firstly, the sum of the rows is always 1, because there are only 4 possible states at any single time for a nucleotide position. Secondly, with no information about the time or r, one could only calculate the distance between two sequences, while the rate or the time could not be calculated. Lastly, as time tends to ∞, with every nucleotide position having had time to change to all four nucleotides, then the probability of observing nucleotide j at each position is ¼. When this happens, we can define the limiting distribution as the proportions of each of the four nucleotides at equilibrium $(\pi_A, \pi_G, \pi_T, \pi_C)$.

What is the proportion of differences between two nucleotide sequences at time $t+1$? Let's assume that the proportion of identical nucleotides between two sequences is $q_{ii}(t)$ at time t, while the proportion of differences between these two sequences is $q_{ij}(t)$ (which is identical to $1 - q_{ii}$). To determine the proportion of differences between the two sequences at time $t + 1$, we need to calculate the two quantities $q_{ii}(t + 1)$ and $q_{ij}(t + 1)$. The probability that two sequences that had the same nucleotide at time t remain to show such state at time $t + 1$ is $(1 - r)^2$. Because the rate of substitutions is very low in an infinitesimal amount of time, r^2 is negligible and thus the solution to this is $q_{ii}(t + 1) = 1 - 2r$. Similarly, a site that had different nucleotides (i for sequence X and j for sequence Y) at time t can show identical nucleotides at time $(t + 1)$ if nucleotide i changes to nucleotide j, while j remains conserved or vice versa. The probability of nucleotide i changing to j (α) and nucleotide j remaining constant (1-r) is therefore:

$$\alpha(1 - r) = \tfrac{r}{3}(1 - r) \tag{3.7}$$

Taking into account the complementary scenario, that is nucleotide i remaining conserved at $(t + 1)$ while j changing to i, then the probability of the sequence remaining conserved at time $(t + 1)$ is $2r(1 - r)/3$. Given that r^2 is approximately 0 at an infinitesimal amount of time this term can be written as $2r/3$. We can now calculate the transition matrix of identity between two sequences at time $(t + 1)$ as:

$$q_{t+1} = (1 - 2r)q_t + \frac{2r}{3}(1 - q_t) \tag{3.8}$$

The variation in the transition matrix (or increment Δq) can be calculated as:

$$\Delta q_t = q_{t+1} - q_t = \frac{2r}{3} - \frac{8r}{3}q_t \tag{3.9}$$

Inferring the amount of continuous change of the matrix along time is possible by integrating (3.9) as:

$$\int \frac{dq}{dt}\,dt = 1 - \frac{3}{4}\left(1 - e^{\frac{-8rt}{3}}\right) \tag{3.10}$$

The expected number of nucleotide substitutions per site (d) for the two sequences being compared is $2rt$. In this case, d is given by

$$d = -\left(\tfrac{3}{4}\right)\ln\left(1 - \left(\tfrac{4}{3}\right)p\right) \tag{3.11}$$

Here p refers to the proportion of nucleotide differences between the two sequences being compared, hence $p = 1 - q$ (Jukes and Cantor 1969). The variance and estimates of p have been previously obtained (Kimura and Ota 1972). For the sake of simplicity, readers should refer to these studies for further details.

At equilibrium, the JC model assumes equal nucleotide frequencies, although we need not assume such a case since we have no knowledge of the initial nucleotide frequencies (Rzhetsky and Nei 1995).

3.2.1.2 The two-parameters model of Kimura

There are two types of nucleotides according to their chemical properties, the purines (A and G) and the pyrimidines (T and C). The transitions between two purines (A <-> G) or two pyrimidines (T <-> C) are more frequent than the transition between a purine and pyrimidine (also known as transversions). The reasons for this asymmetry in the transitions probabilities are manifold: (a) nucleotides within the same group are chemically more similar than between groups. For example, purines are the result of a fusion between six-membered and five-membered nitrogen-containing rings, while pyrimidines result from the fusion of six-membered only nitrogen-containing groups (b) transitions in protein-coding genes often lead to codons that codify the same amino acids (synonymous substitutions), while transversions, especially in two-degenerate and six-degenerate codons such as Leucine, Serine, Arginine, Aspartate, Glutamate, Asparagine and Glutamine, lead to codons codifying different amino acids (non-synonymous substitutions). Because synonymous substitutions often are not seen by Natural selection, such substitutions can become neutrally fixed in the genomes. The opposite is true for non-synonymous substitutions.

In agreement with this rationale, the number of transitions measured from empirical data is often manyfold larger than the number of transversions. Taking into account this observation, Kimura (Kimura 1980) developed a model to estimate the number of nucleotide substitutions based on unequal probabilities for transitions (ts) and transversions (tv). This model also allows estimating explicitly *ts* (symbolized in Kimura's model as α) and *tv* (symbolized as β). Because there are twice as many types of transversions than transitions (Figure 3.1), the total number of nucleotide substitutions can be estimated as α + 2β. Under his model, the total number of transitions (P) and transversions (Q) between two sequences can be estimated as:

$$P = \left(\tfrac{1}{4}\right)\left(1 - 2e^{-4(\alpha+\beta)t} + e^{-8\beta t}\right) \tag{3.12}$$

$$Q = \left(\tfrac{1}{2}\right)\left(1 - e^{-8\beta t}\right) \tag{3.13}$$

Here, t is the time of divergence between the two sequences under comparison. This time can be estimated in different ways, including the number of synonymous substitutions between the two sequences or the length of the branches in the phylogenetic tree leading to the sequence of interest. Accounting for both, *ts* and *tv*, one can estimate the number of nucleotide substitutions between two sequences as being:

$$d = -(1/2)ln(1 - 2P - Q) - (1/4)ln(1 - 2Q) \tag{3.14}$$

The variances associated to each of these estimates have been estimated elsewhere (Nei and Kumar 2000).

3.2.1.3 Complex models of DNA evolution

The two-parameters Kimura and JC69 models assume that the frequencies (π) of all four nucleotides are identical

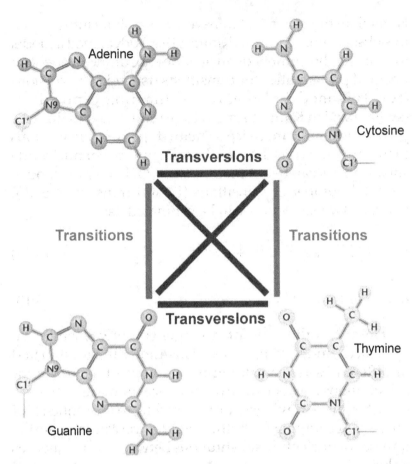

Figure 3.1. Transitions and Transversions in DNA sequences. The number of transitions (red lines) is half as numerous as that of transversions (blue lines). Transitions, on the one hand, refer to substitutions between nucleotides that are chemically similar (Adenine A and Guanidine G or Citosine C and Thimine T). Transversions, on the other hand, stand for substitutions between chemically different nucleotides (A or G and C or T).

Color image of this figure appears in the color plate section at the end of the book.

in stationary phase (in equilibrium), so that $\pi_A = \pi_G = \pi_C = \pi_T = \frac{1}{4}$. In real data, however, the frequency of the four nucleotides is never identical and can vary to different degrees according to the organism's features. For example, bacteria display an astonishing variation in their overall base composition, ranging from 13% to over 75% G+C (McCutcheon and Moran 2010). This variation has been thought to be a neutral character, in which mutational processes vary between species (Freese 1962; Sueoka 1962). Recently, it has been shown that selective forces favor the increase in GC content of bacteria (Raghavan et al. 2012).

The model most used that accounts for variation in the frequency of nucleotides when estimating the nucleotide divergence between two sequences per site is that of Tamura and Nei (TN93:(Tamura and Nei 1993)). In addition to this model, two other models, which are special cases of the TN93 model, have been frequently used: the model of Felsenstein (F83:(Felsenstein 1983)) and the model of Hasegawa, Kishino and Yano (HKY85:(Hasegawa et al. 1985)), both of which assume unequal frequencies of the four nucleotides. Suffice to say that, HKY85 model is a special case of TN93 and it does estimate, in addition to the frequencies of the nucleotides, the transitions and transversions rates. The equations of these models are too complicated to explain here and go beyond the scope of this book. Instead, I will devote the remainder of the chapter to show the main differences in the estimates of nucleotide substitutions per site under the different models and the conditions under which one model is preferable to use against other alternative models (for further reading on the mathematics of the model, I refer readers to Yang 2006).

One complication when estimating the divergence levels, measured as the number of nucleotide substitutions per site, between two species is choosing the appropriate model of nucleotide substitutions. Models are generally differently fitted to explain the divergence levels of particular genome regions and may not be so suitable for other regions—that is, identity in the rates of evolution between nucleotide sites rarely holds. Traditionally, the heterogeneity in the rates of evolution between genomic regions has been ultimately attributed to the interplay between natural selection and genetic drift over a background of universally uniform mutational processes. A growing body of evidence supports mechanistic and compositional reasons for the heterogeneity of substitution rates between coding and non-coding regions and within each of these regions (Subramanian and Kumar 2003; Lercher et al. 2004). Also, the heterogeneity in substitution rates has been linked to the variation in the composition of genomic regions as well as to the chemical differences between the nucleotides (Galtier et al. 2001). Other phenomena that may influence the patterns of variation in evolutionary rates include, among others, the process of replication (Prioleau 2009) and transcription (Mugal et al. 2009), the mutagenesis induced by recombination in mammals (Galtier et al. 2001), differential rates of sex-biased germ line mutation (Nachman and Crowell 2000) and context-dependent effects (Hodgkinson et al. 2009).

Statistically, it has been shown that heterogeneity in the substitution rates among sites follows approximately a gamma distribution (Kocher and Wilson 1991; Tamura and Nei 1993; Wakeley 1993, 1994). There are two parameters that define the gamma distribution: The shape parameter (α) and the scale parameter (β) (Figure 3.2). Of these two parameters, α is the one defining the skew of the distribution; hence the magnitude of heterogeneity in substitution rates among sites. Different nucleotide substitution models have derived the gamma distance. It isn't the aim of this book to present such derivations but it is important to highlight that when the different nucleotide positions follow the same

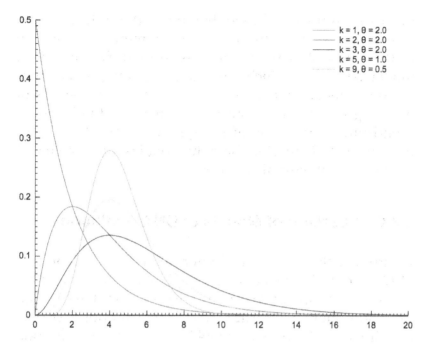

Figure 3.2. Substitution rates distribute according to a gamma distribution. The gamma distribution measures the heterogeneity in the distribution of a parameter across a sample space. In terms of substitution rates and DNA sequences, gamma distribution provides an idea of the heterogeneity of substitution rates across sequences. The shape of the distribution is defined by the shape parameter k, the larger value of this parameter indicating a more homogeneous distribution of the rates of substitutions. The scale of the distribution is given by θ.

Color image of this figure appears in the color plate section at the end of the book.

model of evolution, the rates of evolution of such positions can vary to a greater or lesser extent depending on the value of the shape parameter of the gamma distribution (lower α values means greater heterogeneity and vice versa). Many other parameters define the models of nucleotide substitutions, such as the number of invariable sites in the sequence, the reversibility of nucleotide substitutions, etc. The combinatorial use of these parameters has given birth to a number of models whose fitness to the data can be compared using probabilistic methods. The aim of these probabilistic methods is to identify the best-fit model to the data, which can be subsequently used to infer the phylogenetic relationships between the sequences of homologous genes. Because these trees are the basis for identifying signatures of natural selection, it becomes obvious that identifying the right model is fundamental to infer events of natural selection.

3.3 Comparison of Models of DNA Evolution

Identifying the model that better describes the evolution of DNA data is crucial, as most phylogenetic inferences and identification of selective constraints are based upon the assumption that the model of DNA evolution is correct (Goldman 1993). Models of DNA evolution are built upon a number of parameters, which describe with certain confidence the dynamics of DNA evolution. For example, JC69 model assumes that the transitions between any two nucleotides have the same likelihoods, while K80 model takes into account a greater likelihood for transitions compared to transversions. Other models also account for the direction of the transition and transverion, for the variability of particular nucleotide sites or the transition transversion rates ratio. In theory, the number of possible models can be substantial; with the larger the number of parameters in the model the higher is the likelihood of it being a realistic representation of the evolution of DNA sequences. Very complex models bear, however, the inconvenience of over-fitting—when the number of parameters exceeds the ability of the model to describe the evolution of the data.

A number of tests have been designed to resolve the tradeoff between model complexity and its explanatory power. One such powerful test is the one in which the log-likelihood value of the complex model is compared to that of a more simple model. When the comparison is performed over two nested models—for example, the parameters inferred in the simple model are included within the wider range of parameters inferred by the complex model—then a likelihood ratio test (LRT) can be performed:

$$LRT = -2log\Lambda \qquad (3.15)$$

$$\Lambda = \frac{max[L_0(Simple\ model|Data)]}{max[L_1(Complex\ model|Data)]} \qquad (3.16)$$

LRT can be approximated to a χ^2 distribution, with the degrees of freedom being the difference in the number of parameters estimated in the complex and simple models (Kendall and Stuart 1979; Goldman 1993). For example, in the JC69 model there are three estimated parameters corresponding to the frequencies of the three nucleotides (π_A, π_T, π_C). In the K80 model, the transition to transversion rates ratio, in addition to the nucleotide frequencies, is estimated. Therefore, the log-likelihood value of JC69 (ℓ_0) can be compared to that of K80 (ℓ_1), the more complex model, using LRT and the significance of the difference calculated through a χ^2 distribution with 1 degree of freedom. A significant difference would indicate that K80 yields significantly higher likelihood values than JC69.

While the LRT approximation to a χ^2 seems to work fine, this distribution can be unreliable when the simple model (also the null one) involves fixing the parameters at the boundary of the parameter space of the alternative model—for example, considering a model with homogeneous distribution of substitutions across the sequence as a particular case of a heterogeneous model with the shape parameter of the gamma distribution (α) equal to infinity (Yang 1996). To account for this problem, Ota et al. (Ota et al. 2000) and Goldman and Whelan (Goldman and Whelan 2000) suggested the use of a mixed χ^2 distribution consisting of a 50% χ_0^2 and 50% χ_1^2 to construct a LRT for the invariable sites.

When the models being compared are not nested, then the performance of the model can be compared to that of another by the Akaike information criterion (AIC; (Akaike 1974)) or the Bayesian information criterion (BIC; (Schwarz 1978)). AIC is an unbiased estimator of the Kullback-Leibler information quantity (Kullback and Leibler 1951), which provides a measure of the amount of information lost when the models approximate a more realistic representation of the data by adding more parameters. That is, AIC gives a measure of how over-fit a model is. To account for the likelihood and number of parameters included in the model, AIC is calculated as:

$$AIC_i = -2lnL_i + 2K_i \tag{3.17}$$

where K_i is the number of free parameters in the *i*th model and L_i is the maximum likelihood of the data under the *i*th model.

When the number of characters is small compared to the number of parameters used in the model (and this is an arbitrary qualification decided by the user), then a correction to the AIC can be approximated to:

$$AIC_C = AIC + \frac{2K(K+1)}{n-K-1} \tag{3.18}$$

Generally, the model that minimizes the AIC is the one considered to be the best model representing the data. In addition, the approximation of the model to a true representation of the observed data can be obtained by the weighted AIC for R candidate models:

$$w_i = \frac{exp(-1/2\Delta_i)}{\sum_{r=1}^{R} exp(-1/2\Delta_r)} \qquad (3.19)$$

$$\Delta_i = AIC_i - \min AIC \qquad (3.20)$$

The comparison of the likelihoods of two non-nested models can also be performed using the Bayesian Criterion (BIC). In this case, the size of the data (for example, number of characters in the alignment of the DNA sequences) is incorporated in the comparison:

$$BIC_i = -2lnL_i + K_i \log n \qquad (3.21)$$

K_i and L_i both have the same interpretation as in eq. 3.17, and n is the sample size. The BIC model is directly related to the posterior probability of the model, meaning that the model with the minimum BIC is the one with the maximum posterior probability (Wasserman 2000). The BIC approximation is not always similar to AIC: i) It tends to select simpler models than AIC when n > 8 (Foster and Sober 2004), and ii) BIC is not appropriate when the posterior mode occurs at the boundaries of the parameter space (Hsiao 1997; Ota et al. 2000).

References

Akaike, H. 1974. A new look at the statistical model identifications. Trans Automat Contr 19: 8.
Felsenstein, J. 1983. Statistical inference of phylogenies. Journal of the Royal Statistics Society A 146: 27.

Foster, M. and E. Sober. 2004. How to Tell When Simpler, More Unified, or Less Ad Hoc Theories will Provide More Accurate Predictions. The British Journal for the Philosophy of Science 45: 35.

Freese, E.B. 1962. Further studies on induced reverse mutations of phage T4 rII mutants. Biochemical and Biophysical Research Communications 7: 18–22.

Galtier, N., G. Piganeau, D. Mouchiroud and L. Duret. 2001. GC-content evolution in mammalian genomes: the biased gene conversion hypothesis. Genetics 159: 907–911.

Goldman, N. 1993. Statistical tests of models of DNA substitution. Journal of Molecular Evolution 36: 182–198.

Goldman, N. and S. Whelan. 2000. Statistical tests of gamma-distributed rate heterogeneity in models of sequence evolution in phylogenetics. Molecular Biology and Evolution 17: 975–978.

Gu, X. and W.H. Li. 1998. Estimation of evolutionary distances under stationary and nonstationary models of nucleotide substitution. Proc Natl Acad Sci U S A 95: 5899–5905.

Hasegawa, M., H. Kishino and T. Yano. 1985. Dating of the human-ape splitting by a molecular clock of mitochondrial DNA. Journal of Molecular Evolution 22: 160–174.

Hodgkinson, A., E. Ladoukakis and A. Eyre-Walker. 2009. Cryptic variation in the human mutation rate. PLoS Biology 7: e1000027. doi: 10.1371/journal.pbio.1000027.

Hsiao, C.K. 1997. Approximate Bayes factors when a mode occurs on the boundary. Journal of the American Statistical Association 8.

Jukes, T.H. and C.R. Cantor. 1969. Evolution of protein molecules. H.N. Munro Ed. New York.

Kendall, D.G. and A. Stuart. 1979. The advanced Theory of Statistics. London: Griffin.

Kimura, M. 1980. A simple method for estimating evolutionary rates of base substitutions through comparative studies of nucleotide sequences. Journal of Molecular Evolution 16: 111–120.

Kimura, M. and T. Ota. 1972. Mutation and evolution at the molecular level. Tanpakushitsu kakusan koso Protein, nucleic acid, enzyme 17: 401–413.

Kocher, T.D. and A.C. Wilson. 1991. Sequence evolution of mitochondrial DNA in humans and chimpanzees: Control region and a protein-coding region. Osawa, S. and T. Honjo Eds. Springer Verlag, New York.

Kullback, S. and R.A. Leibler. 1951. On information and sufficiency. Annals of Mathematical Statistics 22: 8.

Lercher, M.J., J.V. Chamary and L.D. Hurst. 2004. Genomic regionality in rates of evolution is not explained by clustering of genes of comparable expression profile. Genome Research 14: 1002–1013. doi: 10.1101/gr.1597404.

McCutcheon, J.P. and N.A. Moran. 2010. Functional convergence in reduced genomes of bacterial symbionts spanning 200 My of evolution. Genome Biology and Evolution 2: 708–718. doi: 10.1093/gbe/evq055.

Mugal, C.F., H.H. von Grunberg and M. Peifer. 2009. Transcription-induced mutational strand bias and its effect on substitution rates in human genes. Molecular Biology and Evolution 26: 131–142. doi: 10.1093/molbev/msn245.

Nachman, M.W. and S.L. Crowell. 2000. Estimate of the mutation rate per nucleotide in humans. Genetics 156: 297–304.

Nei, M. and S. Kumar. 2000. Molecular Evolution and Phylogenetics. Nei, M. and S. Kumar Eds. Oxford University Press, Oxford.

Ota, R., P.J. Waddell, M. Hasegawa, H. Shimodaira and H. Kishino. 2000. Appropriate likelihood ratio tests and marginal distributions for evolutionary tree models with constraints on parameters. Molecular Biology and Evolution 17: 798–803.

Prioleau, M.N. 2009. CpG islands: starting blocks for replication and transcription. PLoS Genetics 5: e1000454. doi: 10.1371/journal.pgen.1000454.

Raghavan, R., D.B. Sloan and H. Ochman. 2012. Antisense transcription is pervasive but rarely conserved in enteric bacteria. mBio 3. doi: 10.1128/mBio.00156-12.

Rzhetsky, A. and M. Nei. 1995. Tests of applicability of several substitution models for DNA sequence data. Molecular Biology and Evolution 12: 131–151.

Schwarz, G.E. 1978. Estimating the dimension of a model. Annals of Statistics 6: 4.

Subramanian, S. and S. Kumar. 2003. Neutral substitutions occur at a faster rate in exons than in noncoding DNA in primate genomes. Genome Research 13: 838–844. doi: 10.1101/gr.1152803.

Sueoka, N. 1962. On the genetic basis of variation and heterogeneity of DNA base composition. Proc Natl Acad Sci U S A 48: 582–592.

Swofford, D.L., G.J. Olsen, P.J. Waddell and D.M. Hillis. 1996. Phylogenetic Inference in Molecular Systematics. Sinauer Associates, Inc. Sunderland.

Tamura, K. and M. Nei. 1993. Estimation of the number of nucleotide substitutions in the control region of mitochondrial DNA in humans and chimpanzees. Molecular Biology and Evolution 10: 512–526.

Wakeley, J. 1993. Substitution rate variation among sites in hypervariable region 1 of human mitochondrial DNA. Journal of Molecular Evolution 37: 613–623.

Wakeley, J. 1994. Substitution-rate variation among sites and the estimation of transition bias. Molecular Biology and Evolution 11: 436–442.

Wasserman, L. 2000. Bayesian model selection and model averaging. Journal of Mathematical Psychology 44: 16.

Yang, Z. 1996. Among-site rate variation and its impact on phylogenetic analyses. Trends in Ecology & Evolution 11: 367–372.

Yang, Z. 2006. Computational Molecular Evolution. Oxford: Oxford University Press.

Zharkikh, A. 1994. Estimation of evolutionary distances between nucleotide sequences. Journal of Molecular Evolution 39: 315–329.

CHAPTER 4

Identifying Natural Selection with Molecular Data

Mario A. Fares

4.1 Introduction

Most methods to identify natural selection using molecular data rely upon the assumption that molecules contribute a relative amount to the biological fitness of individual organisms. This relative contribution is often difficult to quantify, although molecular evolution studies work on the assumption that essential proteins are generally subjected to stronger selective constraints than non-essential proteins. While this view can be challenged by the fact that shifts on selective constraints are possible from one group of organisms to another for a particular protein, essential proteins in one organism are generally so in its close phylogenetic relatives. How could we possibly identify the signatures of natural selection using molecular data? In a previous chapter, we have learned that protein-coding genes comprise sets of codons (nucleotide triplets), in which mutations have different rates of fixation

School of Genetics & Microbiology, Dept. of Genetics, University of Dublin, Trinity College, Dublin 2, Dublin, Ireland.

depending on whether they fall in the first, second or third positions of the codon. Mutations that lead to amino acid replacements, generally those falling in the first and second codon positions, are generally more harmful than the ones that do not change the amino acid. However, the relative importance of amino acid substitutions can differ according to the implication of the affected amino acid in the protein function or structure. Amino acid replacements that do not alter any of the properties of proteins, hence having no effect on the performance of an organism, are known as neutral changes. Those replacements that involve a change in the protein's function are subjected to the filter of natural selection.

Against expectations raised by those belonging to the selectionist school, the elegant neutral theory of molecular evolution (Kimura 1983) states that the majority of the nucleotide substitutions that lead to amino acid replacements and which are fixed in the populations are neutral. The predictions of the neutral theory of molecular evolution have been of enormous importance to understand how novel functions emerge in nature and to device statistical measures that approximately quantify the action of natural selection using molecular data.

4.2 The Neutral and Nearly-neutral Theories of Molecular Evolution

Changes at the molecular level are ultimately responsible for morphological, physiological and behavioral adaptations. To what extent genetic and genomic changes map to phenotypic changes remains controversial. Indeed, the neutral theory claims that most of the observed variation within and between species is the result of the random fixation of mutations with little contribution of natural selection (Kimura 1968; King and Jukes 1969). In agreement with this prediction, the amount of intra-population polymorphism is larger than predicted by the selection model, clearly manifested in the 1970's allozymes studies that revealed substantial polymorphism in allozymes through electrophoresis gels (Lewontin 1991).

At the population level, the contribution of neutral evolution to population's polymorphism can be quantified through the parameter of Heterozygosity (H), which is the probability of randomly sampling different alleles from a population in two independent trials. In theory, H of a real population can be compared to one estimated from an ideal population in which individuals mate randomly and with a constant size (Kimura and Crow 1964). The neutral model also makes predictions for the evolution of proteins. According to these predictions, two kinds of mutations can be found under a scenario of neutral evolution: strongly deleterious mutations, generally purged by purifying selection, and strictly neutral mutations that drift to fixation or alternatively, demise. Neutral mutations will have larger probability of fixation in small populations than in large populations. Wright found, however, that in large populations, the fixation rate of neutral mutations is equivalent to the rate of mutation (Wright 1938). The neutral model of molecular evolution found its strongest support in the molecular clock hypothesis, according to which mutations rate is approximately constant per year (Kimura and Ohta 1971b,a). The assumption that different proteins accept a distinct fraction of neutral mutations, allows explaining the variation in the rates of evolution between proteins in the framework of neutrality.

While strict neutrality assumes two classes of mutations, deleterious and neutral, the nearly neutral model introduced a third class of mutations with slight effects on fitness. This class considered in the nearly neutral theory of molecular evolution, comprised mutations with slight deleterious effects and those with slight advantageous effects, including back mutations and compensatory mutations (Ohta 1972b, 1973; Latter 1975; Ohta and Tachida 1990). Under this model, non-neutral mutations (e.g., advantageous or deleterious) will behave effectively as neutral mutations when the stochastic force of random genetic drift overwhelms the force of selection. This situation is a likely scenario when the effective population sizes are sufficiently low. What is most important, the concept of stochasticity gains credibility against selection in mediating the fixation of new mutations if we take into account the effective population size or, as defined in many theoretical studies, the genetic effective population size. Indeed, natural populations are far from the ideal situation in which all potentially reproductive individuals equally reproduce or in which the population is structured as a continuous in space and time. For instance, in eukaryotes not all individuals produce the same number of offspring. Moreover, the populations of many prokaryotes present patchy structures or undergo strong bottlenecks when transferred between generations. An illustrative example is the case of intra-cellular symbiotic bacteria of insects, in which few bacteria cells are transmitted from the bacteriocytes of the mother to the offspring (Buchner 1965; Baumann et al. 1995a,b). Although most credit has been given to natural selection in the emergence of novel adaptations, the importance of genetic drift in shaping the evolution of genome complexity has begun to be appreciated only in the last decade (Lynch and Conery 2003; Charlesworth and Barton 2004; Daubin and Moran 2004; Lynch 2006; Fernandez and Lynch 2011).

A critical parameter determining the fate of a new mutation is the effective population size (N_e) and the coefficient of selection in favor or against the mutations (s).

When the N_e is high so that $|sN_e| \gg 10$, selection against or in favor of mutations is large. Mutations with slight effects on fitness will be selected against or favored if N_e is sufficiently large. Those mutations for which $|sN_e| \ll 1$, will be effectively neutral and can drift randomly to fixation in the population. The fate of some mutations moving in a narrow range between these values depends on the balance between selection and the stochastic effects of random genetic drift (Ohta and Kimura 1971; Ohta 1973). Since mutations occur through an inherently stochastic process, most mutations are deleterious in the current genome background and environment (Hughes and Friedman 2009). However, many mutations that are deleterious in the current environment may be advantageous in other future environments, and thus, their fixation may underlie the emergence of novel functions in future environments (a phenomenon also known as exaptation). In this sense, processes that enable the fixation of deleterious or slightly deleterious mutations, such as genetic linkage, compensatory mutations and mechanisms of mutational robustness, are important in the emergence of biological complexity.

4.3 Testing the Neutral Theory of Molecular Evolution

Over the last two decades, a number of methods have been developed to identify signatures of natural selection. Underlying such methods is a simple hypothesis testing procedure in which the null hypothesis that supports neutral evolution is tested against an alternative selection hypothesis. Molecular data, and in particular data derived from protein-coding sequences, have been tremendously useful for conducting such comparisons. In her study,

Ohta introduced a measure of the action of selection based upon the quantification of non-synonymous mutations and DNA divergence (Ohta 1972a). Currently, the ratio between non-synonymous and synonymous nucleotide substitutions per site is generally employed to identify the evolutionary model that fits data on the divergence of protein-coding sequences. This is based on the assumption that synonymous substitutions are generally free from selection because they do not produce any change in the amino acid composition of proteins. Therefore, the rate of synonymous substitutions equates the rate of neutral nucleotide substitutions (Miyata and Yasunaga 1980). Indeed, the rate of synonymous substitutions is often equivalent between many genes, although factors such as messenger RNA secondary structures and stability, transcription efficiency, splicing, and DNA secondary structures impose constraints on synonymous sites (Xia 1996; Vinogradov 2003; Chamary and Hurst 2005a,b; Chamary et al. 2006; Hoede et al. 2006; Parmley et al. 2006; Warnecke and Hurst 2007). Moreover, selection on synonymous sites has been shown to strongly depend on translational efficiency (Ikemura 1981, 1985; Sharp et al. 1986; Akashi and Eyre-Walker 1998; Duret and Mouchiroud 2000; Duret 2002; Comeron 2004) and translational robustness (Toft and Fares 2009; Zhou et al. 2009). In general, however, and despite these factors, synonymous substitutions are considered to evolve on average neutrally while non-synonymous changes are on average subject to selection because they involve amino acid replacements. Based on this principle, the past decade has witnessed an astronomical increase in the power of identifying selection using protein-coding sequences through the use of the ratio between the rate of non-synonymous and that of synonymous replacements ($\omega = d_N/d_S$). Values of ω serve to infer if mutations are fixed neutrally ($\omega = 1$), are purged by purifying selection ($\omega < 1$), or are positively selected owing to their beneficial impact on fitness ($\omega > 1$). Indeed, a number of studies identified an inverse relationship between ω and estimates of N_e (Lindblad-Toh et al. 2005; Kosiol et al. 2008; Ellegren 2009). Support to the relationship between ω and N_e comes from the higher rates of molecular evolution (measured as ω) in host-dependent bacteria, which bear low N_e, compared to their free-relative bacterial cousins (Moran 1996; Andersson and Andersson 1999; Wernegreen and Moran 1999; Warnecke and Rocha 2011). The N_e of symbiotic bacteria undergoes strong bottlenecks owing to their vertical transmission between generations. As a result of these bottlenecks, the effect of genetic drift counterbalances that of selection (Figure 4.1). Consequently, the rates of protein evolution are accelerated in symbiotic bacteria (Figure 4.1),

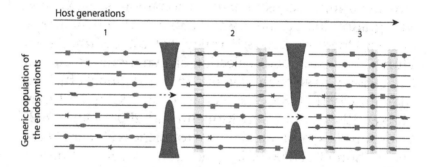

Figure 4.1. Muller's ratchet increases mutational load in endo-cellular symbiotic bacteria. Endo-cellular symbiotic bacteria undergo strong bottleneck during their clonal transmission from one generation to another. Because of these bottlenecks, some mutations (colored symbols), mostly slightly deleterious, drift neutrally to fixation. This leads to a syndrome common to most endo-cellular bacteria transmitted under these regimes: low intra-species polymorphism and high inter-lineage divergence.

Color image of this figure appears in the color plate section at the end of the book.

in particular with regards to the fixation of physico-chemically radical amino acid replacements (Wernegreen 2011). This increase in the rate of evolution mainly occurs between lineages (increase inter-lineage polymorphism) while the polymorphism remains low within lineages.

While nature guards many examples of how mutations can become fixed by genetic drift, the balance between the numbers of synonymous and non-synonymous nucleotide substitutions remains widely used as a rough measure of the strength of natural selection. In the next sections, I will summarize the procedures most generally used to identify selection using either population data or sequences isolated from different species encoding a particular protein.

4.4 Identifying Natural Selection using Population Data: An Introduction

The neutral theory of molecular evolution makes testable predictions, which allows identifying signatures of adaptive evolution. The wealth of sequence data has fuelled the development of a plethora of methods to identify selection, most of which are extensively reviewed in many articles (e.g., Kreitman 2000; Nielsen 2001; Hein et al. 2005). Here, I will only discuss some of the statistics that are and have been widely used to identify natural selection using population data. I will make formal mathematical representations of some of the most used methods, but not of those that are extensions to the former ones. For detailed analytical development of the models underlying the methods, I refer readers to other books (Nielsen 2005; Yang 2006).

4.4.1 Tajima's D statistics

This statistical method, which was first originated by Fumio Tajima (Tajima 1989), measures the amount of genetic variation in a population for a particular set of sequences. There are two types of mutations to distinguish in this test: (a) mutations that segregate in the population at random (also known as polymorphisms, S) and (b) the average proportion of nucleotide differences between all pairs of sequences in the population. S can be estimated using the rate of heterozygosity in the population (θ) and the length in nucleotides of the sequences being compared, generating the estimator of the rate of polymophism in the population ($\hat{\theta}_s$). The average number of pairwise sequence nucleotide differences ($\hat{\theta}_\pi$) can also be estimated from the data. In a randomly mating population, with no selection, recombination or natural selection, the expectation is that $\hat{\theta}_\pi \approx \hat{\theta}_s$. This allows an explicit test of the mode of evolution of sequences in a population by standardizing the comparisons between these two parameters:

$$D = \frac{\hat{\theta}_\pi - \hat{\theta}_S}{SE(\hat{\theta}_\pi - \hat{\theta}_S)} \tag{4.1}$$

Here SE refers to the Standard Error of the differences. Under the null distribution model, D has mean 0 and standard deviation 1. Deviations from these quantities strongly suggest non-neutral evolution, although the alternative hypothesis (evolutionary model) is not straightforward to identify. There are several scenarios under which D may be different from 0. In an expanding population or one in which purifying selection is very strong against new mutations, the number of fixed nucleotide substitutions is low compared to the number of segregating sites, therefore yielding negative D values. Conversely, positive D values can be the result of balancing selection, maintaining genotypes at intermediate levels and can also be the result of population shrinkage.

Similar test statistics to Tajima's D have been explored by other authors, including Fu and Li (Fu and Li 1993) and Fay and Wu (Fay and Wu 2000). Fay and Wu distinguished between mutations occurring at external tree branches and those fixed at internal branches, with the assumption in mind that the former are most likely the product of drift and most such mutations are deleterious or slightly so, while the latter are filtered by natural selection. Under such assumptions, the calculation of D statistic relies on the difference between the estimated number of internal mutations (n_i) and that of external mutations (n_e) as follows:

$$D = \frac{n_i - (a_n - 1)n_e}{SE[n_i - (a_n - 1)n]} \tag{4.2}$$

Here a_n is the summed frequencies of singletons and *SE* refers to the standard error of the differences.

Fay and Wu performed an explicit test of selection by comparing an estimate of θ which takes into account the sizes and number of mutations under two scenarios, one in which sequences evolve neutrally and another in which sequences evolve under selective constraints. The resulting parameter $H = \hat{\theta}_\pi - \hat{\theta}_H$ is expected to have values of 0 when sequences evolve neutrally and negative values otherwise. Obviously, mutations with high frequency contribute substantially to θ_H while those of intermediate frequency contribute more to θ_π.

4.4.2 McDonald-Kreitman test of natural selection

In this test, McDonald and Kreitman (McDonald and Kreitman 1991) developed a test of neutrality under the plausible assumption that mutations observed between species must be considered fixed mutations, whereas those observed within but not between species are not fixed but indicate polymorphism. The expectation is that synonymous and non-synonymous mutations within species should be equal to those between species if such mutations were fixed neutrally. Departure from this prediction would be strongly supporting non-neutral evolution or a complex population dynamics history.

The McDonald-Kreitman test (MKT) is conducted by building contingency 2x2 tables, in which synonymous and non-synonymous mutations are distinguished within and between species. An excess of non-synonymous mutations over polymorphism replacements would indicate positive selection. Conversely, an excess of polymorphisms over mutations between species would be indicative of genetic drift. The difference between the number of polymorphisms and non-synonymous fixed mutations can be tested for significance assuming a χ^2 or Fisher's exact tests.

While this test has proved useful for the analysis of closely related species, it remains very sensitive to the divergence among sequences. Solutions to the problem of multiple hits using several codon models have been provided in subsequent tests (Hasegawa et al. 1998).

Following a similar philosophy, Hudson, Kreitman and Aguadé (Hudson et al. 1987) devised a test of neutral evolution, in which the assumption is that the number of intra-species polymorphisms and inter-species divergence should both be proportional to the mutation rate at a locus. This assumption allows making certain testable predictions on the number of polymorphisms and divergence levels under neutrality. It also sets the ground for verifying such predictions using a test of goodness-of-fit.

4.5 Identifying Natural Selection in Phylogenetic Lineages

Methods to identify natural selection in lineages use the estimation of the number of synonymous substitutions per synonymous site (d_S or K_S) and the number of replacements per non-synonymous site (d_N or K_a) when comparing two or more protein-coding sequences. Since synonymous substitutions make no difference to the protein amino acid composition, they are considered to be on average selectively neutral. It follows then that the ratio ($\omega = d_N/d_S$) is an estimate of the number of amino acids replacing nucleotide substitutions per unit time, a rough estimate of the strength of natural selection. Generally speaking, $\omega = 1$ is an indication of neutral evolution because non-synonymous substitutions evolve at the same rate as synonymous substitutions, implying that such amino acid replacements have no effects on the protein function. Values of $\omega > 1$ or $\omega < 1$ are evidence of positive and negative selection, respectively. While the parameter ω has been widely used to identify signatures of natural selection, caution is required in interpreting results because ω only provides an average measure of the strength of natural selection. Also, the accurate identification of selection is limited by problems underlying the nature of the sequences used. For example, comparison of very divergent sequences means that synonymous sites can be saturated (for example, the weight of multiple hits in synonymous sites is underestimated), thus leading to inflated ω values.

In an attempt to solve these problems, a plethora of models and computational tools have been developed in the last years that provide accurate estimates of the most likely evolutionary scenario giving the data and ω values. These methods could be roughly divided into those that define the most parsimonious evolutionary scenario given the data and those based on probabilistic models that explicitly test *a priori* defined evolutionary hypotheses.

4.5.1 Maximum-likelihood models to identify natural selection

Maximum-likelihood (ML) models assume a Markov model of codon substitution in which the nucleotide or amino acid replacements at time (t) are directly dependent on the previous state in time ($t - 1$). ML models fit this Markov process to the data in hand, namely two protein-coding sequences and a phylogenetic tree, to estimate parameters which are relevant to the evolutionary model proposed, including the transition-transversion rates ratio (κ), ω, the time since the divergence of the two sequences (t), and the frequency of the codon π (Goldman and Yang 1994). The probabilistic nature of ML models makes testing explicit hypotheses possible with regards to the mode of evolution. In general, a model determining a specific evolutionary scenario is tested against a null, simpler model by comparing the likelihoods of both models through the likelihood ratio test (LRT). Calculating the likelihood for a particular model is extensively described in other textbooks (see for instance (Yang 2006)). Comparison of the likelihood between two models, for example a model assuming neutral evolution and another assuming positive selection, is possible using the likelihood ratio test (LRT) as:

$$2\Delta\ell = 2(\ell_0 - \ell_1) \tag{4.3}$$

Here, we compare the log-likelihood of the complex model (ℓ_1) with that of the simpler model (ℓ_0). Twice the difference of likelihoods can be approximated to a χ^2 with the degrees of freedom being the difference in the number of parameters estimated for each of the models. This test can be performed to compare nested models—that is the parameters of the simpler model are a subset of those estimated in the complex model. However, when the models are not nested (e.g., the parameters of the simpler model are not within the set of estimated parameters for the complex model), qualitative ways alternative to the use of LRT allow it comparing the log-likelihoods of the two models, including the Akaike Information Criterion (AIC) (Akaike 1974). AIC is a measure of the relative quality of a model and determines the trade-off between the complexity of the model and the goodness-of-fit of the model. The AIC for a specific model is calculated as:

$$AIC = 2k - 2Ln(\ell) \tag{4.4}$$

Here k refers to the number of parameters and ℓ is the maximized likelihood function. Over-fitted models would involve many more parameters than required to explain the data, thus leading to larger AIC than simpler models.

The number of possible evolutionary scenarios can be considered infinite. The range of interesting models, however, is narrow and limited to those that are of most obvious use and easy interpretation. Current models recreate four general scenarios for the evolution of a protein: (a) the evolution under selection constraints homogeneously distributed across the sequences and lineages of a phylogeny; (b) sequences show equal constraints across the protein but variable along the phylogeny; (c) different regions of a protein differ in their selective constraints, thus evolving at different rates, and (d) the selective constraints on a protein vary between codons and lineages. All these models are implemented in the software codeml from the PAML package version 4.7 (Yang 2007). I will explain some of the most popular models and some particularities with a practical example using PAMLv4.4 (Yang 2007).

The simplest of the models is called G&Y model (Goldman and Yang 1994) and it yields the estimates of a number of important parameters under a model in which a single ω is considered for the protein across the phylogeny and sequence. The parameter ω is estimated from the data and as such it can acquire any value between 0 and infinite. In the majority of cases, this parameter has values below 1, meaning that the protein-coding gene has been evolving on average under purifying selection most of its evolutionary time. For example, I have run a test of evolution in the conserved heat-shock protein GroEL, known to be essential protein to *Escherichia coli* (*E. coli*), and thus expected to evolve under strong purifying selection. The control file of codeml (codeml.ctl) from PAML looks like Figure 4.2. In this control file, there are a number of parameters that need to be specifically examined: (a) the model of evolution (indicated by the options "Model" and "NSsites"); (b) the value of ω needs to be estimated or it can be alternatively fixed by the user; (c) the transition-transvertion rates ratio (Kappa: κ) can be either estimated or fixed; and (d) the shape parameter (alpha: α) of the gamma distribution, which assumes unequal rates of evolution among codons, can also be fixed or estimated from the data.

To run the simplest of the models (G&Y: M0) we need to specify ("Model" = 0 and "NSsites" = 0). Besides from branch lengths, κ and α, the main parameter estimated by G&Y model is that of ω. We can also run the program under a model in which we assume two categories of codon sites, one evolving under strong purifying selection (this category will include codon sites whose ω values will be fixed at $\omega_0 = 0$) and another evolving under strict neutrality (including codons with ω values fixed at $\omega_1 = 1$). The parameters estimated when running this model are those referring to the proportion of codon sites evolving under purifying

```
seqfile = groEL.data        * sequence data filename
treefile = groEL.tree       * tree structure file name
outfile = groEL.out         * main result file name
seqtype = 1                 * 1:codons; 2:AAs; 3:codons-->AAs
CodonFreq = 2               * 0:1/61 each, 1:F1X4, 2:F3X4, 3:codon table
                            *      ndata = 10
model = 0
                        * models for codons:
                        * 0:one, 1:b, 2:2 or more dN/dS ratios for branches

NSsites = 0                     * 0:one w;1:neutral;2:selection; 3:discrete;4:freqs;
                                *5:gamma;6:2gamma;7:beta;8:beta&w;9:beta&gamma
                                *10:beta&gamma+1;11:beta&normal>1;
                                12:0&2normal>1
                        * 13:3normal>0
icode = 0                       * 0:universal code; 1:mammalian mt; 2-10:see below
fix_kappa = 0                   * 1: kappa fixed, 0: kappa to be estimated
kappa = 2                       * initial or fixed kappa
fix_omega = 0                   * 1: omega or omega_1 fixed, 0: estimate
omega = .4                      * initial or fixed omega, for codons or codon-based AAs
fix_alpha = 1                   * 0: estimate gamma shape parameter; 1: fix it at alpha
alpha = 0.                      * initial or fixed alpha, 0:infinity (constant rate)
Malpha = 0                      * different alphas for genes
ncatG = 8                       * # of categories in dG of NSsites models
```

Figure 4.2. Control parameters in the package Phylogenetic Analysis by Maximum Likelihood (Yang 2007). The control file is divided in three main sections. In the first section, essential data concerning the input multiple sequence alignment, phylogenetic tree and output file is provided. This section is followed by one in which the models to be tested are specified. Finally, the last section includes the specification of the parameters to be fixed or inferred by the user for each of the models tested. These parameters include the transition-trasversion rates ratio (Kappa), the non-synonymous-to-synonymous substitutions rates ratio (omega) and the shape parameter of the gamma distribution (alpha).

selection or neutrality. Strictly speaking, these two models, G&Y and Selection-Neutral model (hereafter called M1, and specified by "NSsites" = 1) cannot be directly compared through a LRT as the parameters estimated by the simplest model are not included in the set of parameters estimated by the more complex model. However, we can use the AIC to compare the goodness-of-fit of each model to the data.

The model that minimizes AIC is the one considered to better fit the data. Needless to say that both models, the simpler and more complex models under comparison, may be oversimplifying the real evolutionary scenario and be inappropriate to explain the data.

An alternative scenario to that of purifying and neutral evolution is one in which sequences, parts of the sequences, or particular species of the tree have evolved under adaptive evolution. To test for adaptive evolution operating in a particular protein-coding gene, there are several alternative but not mutually exclusive models. One such model is an extension of the M1 model (PS-M2: "NSsites" = 2) in which an extra category of ω is added and estimated from the data. Under this model, the parameters estimated from the data are ω_2 and the proportion of codon sites classified under two of the ω categories. Due to the nested relationship between M1 and PS-M2, we can compare their log-likelihood values following the LRT. The LRT value is a χ^2 value with the degrees of freedom being the difference in the number of freely estimated parameters in both models (in this case, 1 parameter is estimated for M1 and 3 for PS-M2, and thus *d.f.* = 2).

Another variation to the model of adaptive evolution (PS-M2) is that in which all ω values as well as the proportion of codon sites under each of the categories of ω are estimated from the data. In this model (hereafter called PS-M3: "NSsites" = 3), the problem for many evolutionary biologists when attempting to identify signatures of positive selection is deciding on the number of ω values to be estimated from the data. That is, the number of categories of ω (defined in the codeml.ctl as the "ncatG" option). Intuitively, one may set as many categories as possible (ncatG = 8) because this is the closest we can get to a model of unequal rates among amino acid sites. However, such a configuration of parameters may yield an over-fitting model that may not guarantee a better fit to the data. Instead, one can consider only 3 categories of ω that can account for cases of $\omega < 1$, $\omega = 1$ and $\omega > 1$. This model is comparable to that of G&Y model under the LRT, with the *d.f.* = 5. Alternatives to codon-variation based discrete models are also implemented in PAML, but will not be discussed here (see(Yang and Bielawski 2000)) for further details on the remaining 10 models of evolution implemented in PAMLv4.4.

An alternative model to variation of selective constraints among sites is that assumes differences in the selective constraints among lineages of a phylogeny. As in the case of codons, one can pre-specify a lineage to be tested for presenting a different ω value compared to the remaining lineages of that tree or, alternatively, a model in which each lineage has its own ω value. In either case, the null hypothesis is G&Y model (M0) because this considers equal ω values for the entire tree. To specify a testing branch, the user of PAML can set "Model = 2" (branch model), allowing the test of one or more branches for different ω values. Under this model, the option "NSsites" should remain set to 0. The more complex model (model = 1, also known as the free-ratio model FRM), makes it possible considering a hypothesis in which each branch evolves under different selective constraints in comparison to the rest of the phylogeny. Branch model and FRM can both be compared to the G&Y model using the LRT and with the degrees of freedom being the number of branches for which ω is estimated minus 1.

While these models, branch models and codon models, have substantially improved our ability to identify natural selection compared to previous approaches, they are not free of limitations. For instance, these models are very conservative because natural selection, in particular adaptive evolution, is likely to occur punctually during the evolution of a gene. Branch models allow identifying positive selection ($\omega > 1$) events provided that the selection footprints are sufficiently strong or pervasive in that branch. If ω has been > 1 for a few number of codons but most of the gene has evolved under strong purifying selection ($\omega \approx 0$), then the average ω value for the sequence would be closer to 0 than it would to 1, yielding on average purifying selection for that branch. Similarly, if a gene has evolved under positive selection in one branch but under strong purifying selection in the remaining of the tree, then it would be unlikely to detect positive selection in that gene using branch-based models.

In an attempt to cover these limitations, models that account for the variation of selective constraints at both, branches and codons, have been devised and implemented in PAMLv4.4. For example, Yang and Nielsen (Yang and Nielsen 2002) developed a model in which they divided branches into two categories, foreground and background branches. The model (branch-site model) allows calculating the probability of detecting positive selection in a particular set of codons along the foreground lineages. The log-likelihood value of this model (that the pre-specified foreground branches evolve differently to background branches) is compared with its null hypothesis that assumes no positive selection in the foreground branch.

The branch-site model was shown to be very sensitive to the violation of the underlying assumptions, leading to a significant number of false positives (Zhang 2004). This prompted the development of more robust branch-site models (Yang and Rannala 2005; Yang et al. 2005; Zhang et al. 2005). Other methods have been developed to identify selection signatures at individual codon sites using a maximum-likelihood approach. These methods have been implemented in the server Datamonkey (www.datamonkey. org) and are extensively discussed in Kosakovsky Pond and Frost 2005.

4.5.2 Window-based parsimony approaches to identify natural selection

Models to identify natural selection are based upon the assumption that codon sites in a protein evolve independently from one another. While this assumption may hold for some sites, the chances are high that most sites in a protein depend on one another because of structural and functional ties. Taking this dependency into account may improve our ability to estimate the pervasiveness of adaptive evolution in one protein or to extend the estimates of ω to regions that comprise several interlinked codons.

Two main methods have been developed to account for dependencies among codons and amino acid sites, one that estimates the action of natural selection in statistically determined sequence codon windows (Fares et al. 2002b) and another that examines the selective constraints by sliding windows over a protein's structure (Berglund et al. 2005). The first method is implemented in the program SWAPSCv1.0 (Fares 2004) and follows a very simple procedure (Figure 4.3). Given a multiple protein-coding sequence alignment and a phylogenetic tree, the program estimates the mean ω value for the entire alignment. The non-synonymous and synonymous rates are estimated under the model of Li (Li 1993). These estimates take into account the degeneracy of codon sites, transitions and transversions. The program then uses a set of multiple

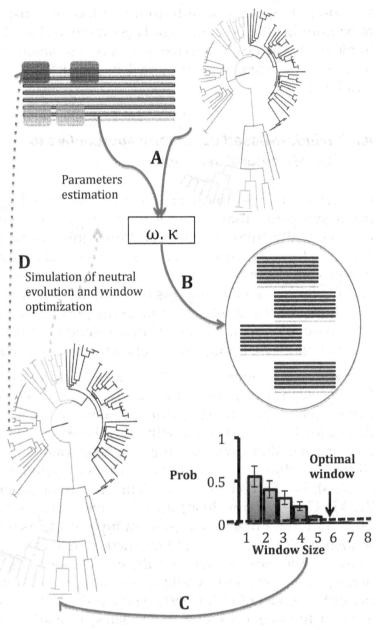

Parameters
estimation

ω, κ

A

B

D
Simulation of neutral
evolution and window
optimization

Prob

Optimal
window

Window Size

C

Figure 4.3. contd....

sequence alignments simulated by other programs under the condition of neutral evolution to statistically optimize the window size required to identify constraints on the real alignment. Sequences are inferred at the ancestral nodes of the tree and d_N and d_S estimated for the comparison of an ancestral node to its child nodes to infer the strength of natural selection along that branch. Since the parameter of selection ω is estimated by sliding a window along the pairwise comparisons, tests of natural selection can be explicitly performed for particular regions of the protein-coding sequences along a tree branch.

The method of tertiary windows by Berglund and colleagues follows the same philosophical methodology as the one in SWAPSC, although it utilizes tertiary windows slid along the crystal structure of a protein such that correlated evolution may not necessarily occur only among linearly neighbor amino acid sites but also at structurally neighbor sites (Berglund et al. 2005). While this method is furnished with the pragmatism of a more realistic biological scenario than that in SWAPSC, it suffers from the obvious limitation of the lack of three-dimensional information for the majority of known proteins.

Figure 4.3. contd.

Figure 4.3. Flow chart of the software Sliding Window Analysis using Protein Sequences (SWAPSC). Multiple sequence alignments (MSAs) and the phylogenetic tree are fed into SWAPSC (A). The non-synonymous-to-synonymous rates ratio (ω) and the transition-transversion rates ratio (κ) are estimated using the alignment and the tree and these parameters are used to simulate MSAs under a neutral evolution model (B). This simulated data are used to infer the size of the window that when slid along the real MSA would provide no significant ω, non-synonymous or synonymous estimates. SWAPSC slides the window in each branch of the tree (C) and window size optimized to provide statistical evidence of selective constraints (D).

Color image of this figure appears in the color plate section at the end of the book.

4.6 The Evolution of GroEL: A Case Study

GroEL is a heat-shock protein (HSP) that belongs to the highly conserved protein family Hsp60 (also known as cpn60). This protein performs, in conjunction with its co-operon encoded cochaperone GroES, essential functions in the cell under normal physiological conditions, chief among which is the folding of a large fraction of cellular proteins. To do so, GroEL interacts with solvent accessible hydrophobic patches of partially unfolded proteins (Braig et al. 1994; Thirumalai and Lorimer 2001; Facciponte et al. 2005; Ellis 2007). The generalist folding function of GroEL is non-specific with regards to protein folding, meaning that proteins compete for binding to GroEL, with those offering the largest hydrophobic surfaces to the solvent being the most GroEL-preferred clients. Under heat-stress, many proteins lose their tertiary folding, thereby exposing cells to the risk of non-specific aggregates. The toxic effects of protein misfolding under heat-stress are, however, counteracted by over-expression of both GroES and GroEL.

Mounting evidence supports other folding- unrelated roles for GroEL, such as immune response in humans (Ronaghy et al. 2011; Zonneveld-Huijssoon et al. 2011) or growth and biofilm formation in bacteria, among others (Ojha et al. 2005; Rodriguez-Quinones et al. 2005; Bittner et al. 2007; Gould et al. 2007; Li et al. 2010; Wang et al. 2013). In summary, GroESL contributes very significantly to the biological fitness of the bacterium and, consequently, changes in this molecule are likely to be swiftly filtered by selection because they directly affect the performance of the cell. In addition to its role in heat-stress response, GroEL has been shown to provide mutational robustness by rescuing mutant proteins with folding defects (Fares et al. 2002a; Fares et al. 2002c; Fares et al. 2004, 2005; Bogumil and Dagan 2010; Williams and Fares 2010). It is hence central to the function of the cell, but it also fuels evolution by allowing the fixation of mutations in proteins requiring GroEL for folding. GroEL has also suffered a remarkable evolution among bacterial groups, showing very distinct and unrelated functions that evolved after GroEL gene duplication or after a change in the lifestyle of the bacterium (see (Henderson et al. 2013) for a recent review).

Due to its essentiality and functional plasticity, GroEL is a good candidate for studying the dynamics of natural selection. To identify selection in *groEL* I have followed the steps:

a) Identify all *groEL* homologs in endosymbiotic and free-living bacteria.
b) Align the amino acid sequences of *groEL* using Clustal Omega (Sievers et al. 2011).
c) Align codons according to the amino acid alignments, so that codon sequences are kept in frame.
d) Infer the phylogenetic tree by maximum-likelihood for *groEL* homologous sequences using PhyML, and support internal nodes of the tree using bootstrap (Guindon et al. 2010).
e) Finally, test models of evolution of codon-sequences for goodness-of-fit to the data using the package PAML4.7 (Yang 2007).

Alignment of sequences and phylogenetic tree inference software are integrated within the package SeaView for Mac OS X Version 4.4.0 (Gouy et al. 2010).

The phylogeny inferred by PhyML is illustrated in Figure 4.4 and represents a maximum-likelihood phylogenetic tree for bacteria, which established a mutualistic association with aphids and their free-living bacterial cousins. Bacteria

71

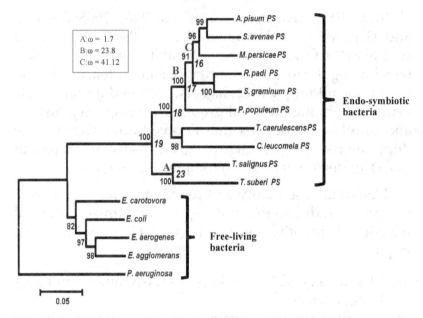

Figure 4.4. Identification of adaptive evolution in the heat-shock protein GroEL of endosymbiotic bacteria of aphids. We applied the program SWAPSC to identify adaptive evolution in endosymbiotic bacteria of aphids. Three of the branches showed non-synonymous-to-synonymous rates ratio (ω) greater than 1 (branches A to C).

Color image of this figure appears in the color plate section at the end of the book.

with different lifestyles to their original ancestor may have undergone selective shifts in key proteins, with GroEL being a good candidate. To test this hypothesis, we set the file "codeml.ctl" from the program CODEML control file as shown in Figure 4.2. As main files, codeml.ctl contains input (groEL.data) and output (groEL.out) files as well as the file containing the phylogenetic tree in newick format (groEL.tree). I searched for signatures of positive Darwinian selection in this file. To do so, one can run a model in which all codon sites of the alignment evolve under the same model (NSsites = 0) and evolution does not change along

72

the phylogenetic tree, meaning that a single ω is estimated for the entire alignment and phylogeny (Model = 0). The parameters are estimated directly from the data:

a) The transition-to-transvertion rates ratio (Kappa) is estimated (fix kappa = 0) departing from an initial given value (Kappa = 2).
b) The ω value is also estimated from the data (fixed omega = 0), seeding at a given ω value (gamma = 0.4).

All the other parameters remain unchanged, as they do not directly apply to the model in use or have little influence on the results.

The null hypothesis (NSsites = 0, Model = 0; G&Y) is compared to an alternative one in which branches are allowed to evolve freely (Free-ratio model: FRM), and thus a ω value is estimated for each of the lineages of the tree. To run this model, one can set the parameter (Model = 1) and (NSsites = 0). If *groEL* had suffered important adaptive changes in the endosymbiotic clade, then one would expect finding signatures of adaptive evolution within this clade but not in their free-living bacterial sister clade. It is also convenient to conduct this analysis for each of the three functional domains of GroEL (Figure 4.4), namely the apical domain, equatorial domain and intermediate domain. These three domains have distinct biological functions, and thus should be analyzed independently. The results of analyzing these domains are shown in Table 4.1. In terms of parameters, the G&Y model is nested within the FRM and hence their log-likelihoods can be statistically compared using the likelihood-ratio test (LRT). Following this, LRT for the entire protein and for each of the protein domains is shown in Table 4.1.

Table 4.1. Testing Positive selection in GroEL.

GroEL region	l_{FRM}	$l_{G&Y}$	$2\Delta l$	P
Complete	−9544.159	−9637.141	185.964	< 0.001
Apical	−3340.015	−3743.481	802.931	< 0.001
Intermediate	−1573.601	−1597.456	47.710	< 0.001
Equatorial	−4709.947	−5126.318	832.000	< 0.001

All the tests conducted confirmed that the alternative hypothesis (FRM) fits better the data better than the null one (G&Y). The main conclusion of this first analysis is that the different branches of the tree evolved under different selective constraints. Importantly, the FRM provides signatures of positive selection ($\omega > 1$) in three branches of the tree (Figure 4.4), all of which belong to the group of endo-symbiotic bacteria. Also, the output file generated by CODEML provides a list of candidate positively selected codon sites whose probability of belonging to a category of codons with ($\omega > 1$) is higher than 0.95. A number of sites identified under positive selection (for example, they are detected with probability to have fixed amino acid replacements in the lineage with $\omega > 1$), G336D, E340N, H344N, Q346K and E355Q, belong to the region binding unfolded polypeptides and all show radical amino acid replacements. Noticeably, these sites fall close to one another in the protein structure (Figure 4.5 of structure), hinting their coadaptive relationship, probably the result of their functional link.

SWAPSC identified a window of four codons as being statistically optimum to identify selective constraints. The analysis of GroEL using sliding windows approach (SWAPSC) identifies additional regions to those detected by PAML under adaptive evolution in the endosymbiotic bacteria lineages (Table 4.2; full description of the results

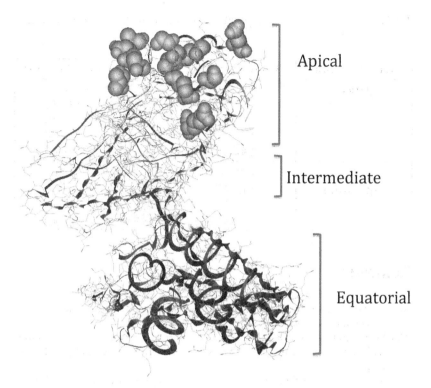

Apical

Intermediate

Equatorial

Figure 4.5. Three-dimensional structure of a GroEL subunit. Spheres labeled amino acid sites identified to have evolved under adaptive evolution in symbiotic bacteria of aphids. All amino acids are clustered in the apical domain of GroEL at sites involved in binding unfolded proteins and the cofactor GroES.

Color image of this figure appears in the color plate section at the end of the book.

are available in Fares et al. 2002a). Amino acid sites identified under positive selection are of importance to protein binding and folding (Figure 4.5 with functional sites marked). Taken together, results point to strong signatures of adaptive evolution in functionally important amino acid sites within the endosymbiotic GroEL.

Table 4.2. Identification of positive selection in the GroEL domains using SWAPSC.

GroEL region	Codon interval	ω	One-tailed P
Apical	246–249	3.53	< 0.001
	339–342	7.20	< 0.001
	344–347	3.52	< 0.001
	348–351	3.54	< 0.001
	350–353	3.58	< 0.001
Intermediate	160–163	4.28	< 0.001
Equatorial	434–437	3.88	< 0.001
	534–537	6.25	< 0.001
	543–546	33.94	< 0.001

References

Akaike, H. 1974. A new look at the statistical model identifications. Trans Automat Contr 19: 8.

Akashi, H. and A. Eyre-Walker. 1998. Translational selection and molecular evolution. Curr Opin Genet Dev 8: 688–693. doi: S0959-437X(98)80038-5 [pii].

Andersson, J.O. and S.G. Andersson. 1999. Insights into the evolutionary process of genome degradation. Curr Opin Genet Dev 9: 664–671. doi: S0959-437X(99)00024-6 [pii].

Baumann, P., L. Baumann, C.Y. Lai, D. Rouhbakhsh, N.A. Moran and M.A. Clark. 1995a. Genetics, physiology, and evolutionary relationships of the genus Buchnera: intracellular symbionts of aphids. Annual Review of Microbiology 49: 55–94. doi: 10.1146/annurev.mi.49.100195.000415.

Baumann, P., C. Lai, L. Baumann, D. Rouhbakhsh, N.A. Moran and M.A. Clark. 1995b. Mutualistic associations of aphids and prokaryotes: biology of the genus buchnera. Applied and Environmental Microbiology 61: 1–7.

Berglund, A.C., B. Wallner, A. Elofsson and D.A. Liberles. 2005. Tertiary windowing to detect positive diversifying selection. J Mol Evol 60: 499–504. doi: 10.1007/s00239-004-0223-4.

Bittner, A.N., A. Foltz and V. Oke. 2007. Only one of five groEL genes is required for viability and successful symbiosis in Sinorhizobium meliloti. J Bacteriol 189: 1884–1889. doi: 10.1128/JB.01542-06.

Bogumil, D. and T. Dagan. 2010. Chaperonin-dependent accelerated substitution rates in prokaryotes. Genome Biol Evol 2: 602–608. doi: 10.1093/gbe/evq044.

Braig, K., Z. Otwinowski, R. Hegde, D.C. Boisvert, A. Joachimiak, A.L. Horwich and P.B. Sigler. 1994. The crystal structure of the bacterial chaperonin GroEL at 2.8 A. Nature 371: 578–586. doi: 10.1038/371578a0.

Buchner, P. 1965. Endosymbiosis of animals with plants microorganisms. New York: Interscience Publishers.

Chamary, J.V. and L.D, Hurst. 2005a. Biased codon usage near intron-exon junctions: selection on splicing enhancers, splice-site recognition or something else? Trends Genet 21: 256–259. doi: 10.1016/j.tig.2005.03.001S0168-9525(05)00061-2 [pii].

Chamary, J.V. and L.D. Hurst. 2005b. Evidence for selection on synonymous mutations affecting stability of mRNA secondary structure in mammals. Genome Biol 6: R75. doi: 10.1186/qb-2005-6-9-r75.

Chamary, J.V., J.L. Parmley and L.D. Hurst. 2006. Hearing silence: non-neutral evolution at synonymous sites in mammals. Nat Rev Genet 7: 98–108. doi: 10.1038/nrg1770.

Charlesworth, B. and N. Barton. 2004. Genome size: does bigger mean worse? Curr Biol 14: R233–235. doi: 10.1016/j.cub.2004.02.054.

Comeron, J.M. 2004. Selective and mutational patterns associated with gene expression in humans: influences on synonymous composition and intron presence. Genetics 167: 1293–1304. doi: 10.1534/genetics.104.026351.

Daubin, V. and N.A. Moran. 2004. Comment on "The origins of genome complexity". Science 306: 978; author reply 978. doi: 10.1126/science.1098469.

Duret, L. 2002. Evolution of synonymous codon usage in metazoans. Curr Opin Genet Dev 12: 640–649. doi: S0959437X02003532 [pii].

Duret, L. and D. Mouchiroud. 2000. Determinants of substitution rates in mammalian genes: expression pattern affects selection intensity but not mutation rate. Mol Biol Evol 17: 68–74.

Ellegren, H. 2009. A selection model of molecular evolution incorporating the effective population size. Evolution 63: 301–305. doi: 10.1111/j.1558-5646.2008.00560.x.

Ellis, R.J. 2007. Protein misassembly: macromolecular crowding and molecular chaperones. Adv Exp Med Biol 594: 1–13. doi: 10.1007/978-0-387-39975-1_1.

Facciponte, J.G., I.J. MacDonald, X.Y. Wang, H. Kim, M.H. Manjili and J.R. Subjeck. 2005. Heat shock proteins and scavenger receptors: role in adaptive immune responses. Immunol Invest 34: 325–342.

Fares, M.A. 2004. SWAPSC: sliding window analysis procedure to detect selective constraints. Bioinformatics 20: 2867–2868. doi: 10.1093/bioinformatics/bth329.

Fares, M.A., E. Barrio, B. Sabater-Munoz and A. Moya. 2002a. The evolution of the heat-shock protein GroEL from Buchnera, the primary endosymbiont of aphids, is governed by positive selection. Mol Biol Evol 19: 1162–1170.

Fares, M.A., S.F. Elena, J. Ortiz, A. Moya and E. Barrio. 2002b. A sliding window-based method to detect selective constraints in protein-coding genes and its application to RNA viruses. J Mol Evol 55: 509–521. doi: 10.1007/s00239-002-2346-9.

Fares, M.A., M.X. Ruiz-Gonzalez, A. Moya, S.F. Elena and E. Barrio. 2002c. Endosymbiotic bacteria: groEL buffers against deleterious mutations. Nature 417: 398. doi: 10.1038/417398a.

Fares, M.A., A. Moya and E. Barrio. 2004. GroEL and the maintenance of bacterial endosymbiosis. Trends Genet 20: 413–416. doi: 10.1016/j.tig.2004.07.001.

Fares, M.A., A. Moya and E. Barrio. 2005. Adaptive evolution in GroEL from distantly related endosymbiotic bacteria of insects. Journal of Evolutionary Biology 18: 651–660. doi: 10.1111/j.1420-9101.2004.00861.x.

Fay, J.C. and C.I. Wu. 2000. Hitchhiking under positive Darwinian selection. Genetics 155: 1405–1413.

Fernandez, A. and M. Lynch. 2011. Non-adaptive origins of interactome complexity. Nature 474: 502–505. doi: 10.1038/nature09992.

Fu, Y.X. and W.H. Li. 1993. Statistical tests of neutrality of mutations. Genetics 133: 693–709.

Goldman, N. and Z. Yang. 1994. A codon-based model of nucleotide substitution for protein-coding DNA sequences. Mol Biol Evol 11: 725–736.

Gould, P., M. Maguire and P.A. Lund. 2007. Distinct mechanisms regulate expression of the two major groEL homologues in Rhizobium leguminosarum. Archives of Microbiology 187: 1–14. doi: 10.1007/s00203-006-0164-y.

Gouy, M., S. Guindon and O. Gascuel. 2010. SeaView version 4: A multiplatform graphical user interface for sequence alignment and phylogenetic tree building. Mol Biol Evol 27: 221–224. doi: 10.1093/molbev/msp259.

Guindon, S., J.F. Dufayard, V. Lefort, M. Anisimova, W. Hordijk and O. Gascuel. 2010. New algorithms and methods to estimate maximum-likelihood phylogenies: assessing the performance of PhyML 3.0. Systematic Biology 59: 307–321. doi: 10.1093/sysbio/syq010.

Hasegawa, M., Y. Cao and Z. Yang. 1998. Preponderance of slightly deleterious polymorphism in mitochondrial DNA: nonsynonymous/synonymous rate ratio is much higher within species than between species. Mol Biol Evol 15: 1499–1505.

Hein, J., M.H. Schierup and C. Wiuf. 2005. Gene genealogies, Variation and Evolution: a primer in coalescent theory. Oxford: Oxford University Press.

Henderson, B., M.A. Fares and P.A. Lund. 2013. Chaperonin 60: a paradoxical, evolutionarily conserved protein family with multiple moonlighting functions. Biological reviews of the Cambridge Philosophical Society 88: 955–987. doi: 10.1111/brv.12037.

Hoede, C., E. Denamur and O. Tenaillon. 2006. Selection acts on DNA secondary structures to decrease transcriptional mutagenesis. PLoS Genet 2: e176. doi: 10.1371/journal.pgen.0020176.

Hudson, R.R., M. Kreitman and M. Aguadé. 1987. A test of neutral molecular evolution based on nucleotide data. Genetics 116: 153–159.

Hughes, A.L. and R. Friedman. 2009. More radical amino acid replacements in primates than in rodents: support for the evolutionary role of effective population size. Gene 440: 50–56. doi: 10.1016/j.gene.2009.03.012.

Ikemura, T. 1981. Correlation between the abundance of Escherichia coli transfer RNAs and the occurrence of the respective codons in its protein genes: a proposal for a synonymous codon choice that is optimal for the *E. coli* translational system. J Mol Biol 151: 389–409. doi: 0022-2836(81)90003-6 [pii].

Ikemura, T. 1985. Codon usage and tRNA content in unicellular and multicellular organisms. Mol Biol Evol 2: 13–34.

Kimura, M. 1968. Evolutionary rate at the molecular level. Nature 217: 624–626.

Kimura, M. 1983. The neutral theory of molecular evolution. Cambridge University Press, London.

Kimura, M. and J.F. Crow. 1964. The Number of Alleles That Can Be Maintained in a Finite Population. Genetics 49: 725–738.

Kimura, M. and T. Ohta. 1971a. On the rate of molecular evolution. J Mol Evol 1: 1–17.

Kimura, M. and T. Ohta. 1971b. Protein polymorphism as a phase of molecular evolution. Nature 229: 467–469.

King, J.L. and T.H. Jukes. 1969. Non-Darwinian evolution. Science 164: 788–798.

Kosakovsky Pond, S.L. and S.D. Frost. 2005. Not so different after all: a comparison of methods for detecting amino acid sites under selection. Mol Biol Evol 22: 1208–1222. doi: 10.1093/molbev/msi105.

Kosiol, C., T. Vinar, R.R. da Fonseca, M.J. Hubisz, C.D. Bustamante, R. Nielsen and A. Siepel. 2008. Patterns of positive selection in six Mammalian genomes. PLoS Genet 4: e1000144. doi: 10.1371/journal.pgen.1000144.

Kreitman, M. 2000. Methods to detect selection in populations with applications to the human. Annu Rev Genomics Hum Genet 1: 539–559. doi:10.1146/annurev.genom.1.1.539.

Latter, B.D. 1975. Enzyme polymorphisms: gene frequency distributions with mutation and selection for optimal activity. Genetics 79: 325–331.

Lewontin, R.C. 1991. Twenty-five years ago in Genetics: electrophoresis in the development of evolutionary genetics: milestone or millstone? Genetics 128: 657–662.

Li, J., Y. Wang, C.Y. Zhang, W.Y. Zhang, D.M. Jiang, Z.H. Wu, H. Liu and Y.Z. Li. 2010. Myxococcus xanthus viability depends on groEL supplied by either of two genes, but the paralogs have different functions during heat shock, predation, and development. J Bacteriol 192: 1875–1881. doi: 10.1128/JB.01458-09.

Li, W.H. 1993. Unbiased estimation of the rates of synonymous and nonsynonymous substitution. J Mol Evol 36: 96–99.

Lindblad-Toh, K., C.M. Wade, T.S. Mikkelsen et al. 2005. Genome sequence, comparative analysis and haplotype structure of the domestic dog. Nature 438: 803–819. doi: 10.1038/nature04338.

Lynch, M. 2006. Streamlining and simplification of microbial genome architecture. Annual Review of Microbiology 60: 327–349. doi: 10.1146/annurev.micro.60.080805.142300.

Lynch, M. and J.S. Conery. 2003. The origins of genome complexity. Science 302: 1401–1404. doi: 10.1126/science.1089370.

McDonald, J.H. and M. Kreitman. 1991. Adaptive protein evolution at the Adh locus in Drosophila. Nature 351: 652–654. doi: 10.1038/351652a0.

Miyata, T. and T. Yasunaga. 1980. Molecular evolution of mRNA: a method for estimating evolutionary rates of synonymous and amino acid substitutions from homologous nucleotide sequences and its application. J Mol Evol 16: 23–36.

Moran, N.A. 1996. Accelerated evolution and Muller's rachet in endosymbiotic bacteria. Proc Natl Acad Sci U S A 93: 2873–2878.

Nielsen, R. 2001. Statistical tests of selective neutrality in the age of genomics. Heredity (Edinb) 86: 641–647. doi: her895 [pii].

Nielsen, R. 2005. Statistical methods in molecular evolution: Springer.

Ohta, T. 1972a. Evolutionary rate of cistrons and DNA divergence. Journal of Molecular Evolution 1: 8.

Ohta, T. 1972b. Population size and rate of evolution. J Mol Evol 1: 305–314.

Ohta, T. 1973. Slightly deleterious mutant substitutions in evolution. Nature 246: 96–98.

Ohta, T. and M. Kimura. 1971. On the constancy of the evolutionary rate of cistrons. J Mol Evol 1: 8.

Ohta, T. and H. Tachida. 1990. Theoretical study of near neutrality. I. Heterozygosity and rate of mutant substitution. Genetics 126: 219–229.

Ojha, A., M. Anand, A. Bhatt, L. Kremer, W.R. Jacobs, Jr. and G.F. Hatfull. 2005. GroEL1: a dedicated chaperone involved in mycolic acid biosynthesis during biofilm formation in mycobacteria. Cell 123: 861–873. doi: 10.1016/j.cell.2005.09.012.

Parmley, J.L., J.V. Chamary and L.D. Hurst. 2006. Evidence for purifying selection against synonymous mutations in mammalian exonic splicing enhancers. Mol Biol Evol 23: 301–309. doi: 10.1093/molbev/msj035.

Rodriguez-Quinones, F., M. Maguire, E.J. Wallington, P.S. Gould, V. Yerko, J.A. Downie and P.A. Lund. 2005. Two of the three groEL homologues in Rhizobium leguminosarum are dispensable for normal growth. Archives of Microbiology 183: 253–265. doi: 10.1007/s00203-005-0768-7.

Ronaghy, A., W. de Jager, E. Zonneveld-Huijssoon, M.R. Klein, F. van Wijk, G.T. Rijkers, W. Kuis, N.M. Wulffraat and B.J. Prakken. 2011. Vaccination leads to an aberrant FOXP3 T-cell response in non-remitting juvenile idiopathic arthritis. Annals of the Rheumatic Diseases 70: 2037–2043. doi: 10.1136/ard.2010.145151.

Sharp, P.M., T.M. Tuohy and K.R. Mosurski. 1986. Codon usage in yeast: cluster analysis clearly differentiates highly and lowly expressed genes. Nucleic Acids Res 14: 5125–5143.

Sievers, F., A. Wilm, D. Dineen et al. 2011. Fast, scalable generation of high-quality protein multiple sequence alignments using Clustal Omega. Mol Syst Biol 7: 539. doi: 10.1038/msb.2011.75.

Tajima, F. 1989. Statistical method for testing the neutral mutation hypothesis by DNA polymorphism. Genetics 123: 585–595.

Thirumalai, D. and G.H. Lorimer. 2001. Chaperonin-mediated protein folding. Annu Rev Biophys Biomol Struct 30: 245–269. doi: 10.1146/annurev.biophys.30.1.245.

Toft, C. and M.A. Fares. 2009. Selection for translational robustness in Buchnera aphidicola, endosymbiotic bacteria of aphids. Mol Biol Evol 26: 743–751. doi: 10.1093/molbev/msn301.

Vinogradov, A.E. 2003. DNA helix: the importance of being GC-rich. Nucleic Acids Res 31: 1838–1844.

Wang, Y., W.Y. Zhang, Z. Zhang et al. 2013. Mechanisms involved in the functional divergence of duplicated GroEL chaperonins in Myxococcus xanthus DK1622. PLoS Genetics 9: e1003306. doi: 10.1371/journal.pgen.1003306.

Warnecke, T. and L.D. Hurst. 2007. Evidence for a trade-off between translational efficiency and splicing regulation in determining synonymous codon usage in Drosophila melanogaster. Mol Biol Evol 24: 2755–2762. doi: 10.1093/molbev/msm210.

Warnecke, T. and E.P. Rocha. 2011. Function-specific accelerations in rates of sequence evolution suggest predictable epistatic responses to reduced effective population size. Mol Biol Evol 28: 2339–2349. doi: 10.1093/molbev/msr054.

Wernegreen, J.J. 2011. Reduced selective constraint in endosymbionts: elevation in radical amino acid replacements occurs genome-wide. PLoS One 6: e28905. doi: 10.1371/journal.pone.0028905.

Wernegreen, J.J. and N.A. Moran. 1999. Evidence for genetic drift in endosymbionts (Buchnera): analyses of protein-coding genes. Mol Biol Evol 16: 83–97.

Williams, T.A. and M.A. Fares. 2010. The effect of chaperonin buffering on protein evolution. Genome Biol Evol 2: 609–619. doi: 10.1093/gbe/evq045.

Wright, S. 1938. The Distribution of Gene Frequencies Under Irreversible Mutation. Proc Natl Acad Sci U S A 24: 253–259.

Xia, X. 1996. Maximizing transcription efficiency causes codon usage bias. Genetics 144: 1309–1320.

Yang, Z. 2006. Computational Molecular Evolution. Oxford: Oxford University Press.

Yang, Z. 2007. PAML 4: phylogenetic analysis by maximum likelihood. Mol Biol Evol 24: 1586–1591. doi: 10.1093/molbev/msm088.

Yang, Z. and J.P. Bielawski. 2000. Statistical methods for detecting molecular adaptation. Trends Ecol Evol 15: 496–503. doi: S0169534700019947 [pii].

Yang, Z. and R. Nielsen. 2002. Codon-substitution models for detecting molecular adaptation at individual sites along specific lineages. Mol Biol Evol 19: 908–917.

Yang, Z. and B. Rannala. 2005. Branch-length prior influences Bayesian posterior probability of phylogeny. Syst Biol 54: 455–470. doi: 10.1080/10635150590945313.

Yang, Z., W.S. Wong and R. Nielsen. 2005. Bayes empirical bayes inference of amino acid sites under positive selection. Mol Biol Evol 22: 1107–1118. doi: 10.1093/molbev/msi097.

Zhang, J. 2004. Frequent false detection of positive selection by the likelihood method with branch-site models. Mol Biol Evol 21: 1332–1339. doi: 10.1093/molbev/msh117.

Zhang, J., R. Nielsen and Z. Yang. 2005. Evaluation of an improved branch-site likelihood method for detecting positive selection at the molecular level. Mol Biol Evol 22: 2472–2479. doi: 10.1093/molbev/msi237.

Zhou, T., M. Weems and C.O. Wilke. 2009. Translationally optimal codons associate with structurally sensitive sites in proteins. Mol Biol Evol 26: 1571–1580. doi: 10.1093/molbev/msp070.

Zonneveld-Huijssoon, E., S.T. Roord, W. de Jager, M. Klein, S. Albani, S.M. Anderton, W. Kuis, F. van Wijk and B.J. Prakken. 2011. Bystander suppression of experimental arthritis by nasal administration of a heat shock protein peptide. Annals of the Rheumatic Diseases 70: 2199–2206. doi: 10.1136/ard.2010.136994.

CHAPTER 5

Inferring Functional Divergence in Protein Sequences

Mario A. Fares[1], and *Christina Toft[2]*

5.1 Introduction

Under the contemporary molecular view of the principle of descendent with modification by Charles Darwin, new genes and functions originate by the modification of ancestral ones, a process known as functional divergence. Both genetic and environmental factors influence the emergence of novel functions. In this chapter we will focus mainly on the methods that allow inferring functional divergence using protein-coding sequences. The underlying assumption when inferring functional divergence from sequences is that the more identical two protein sequences are the more similar their functions are. That is, the divergence

[1] School of Genetics & Microbiology, Dept. of Genetics, University of Dublin, Trinity College, Dublin 2, Dublin, Ireland.
[2] Department of Genetics, University of Valencia, Valencia, Spain; and Instituto de Agroquímica y Tecnología de los Alimentos (I.A.T.A-C.S.I.C).
* Corresponding author

between any two sequences from a single common ancestral sequence is proportional to the divergence in the functions of the proteins they encode. While this assumption may hold true in general, the proportionality between sequence divergence and functional divergence remains debated.

The divergence of a gene towards an alternative function is constrained by the detrimental fitness effect of this functional departure from an ancestral, well adapted, function. Selection generally acts against functional divergence, in particular on those genes that contribute substantially to the biological fitness of an individual. There are two main processes that can, however, relax these constraints on genes, allowing the diversification of functions: gene duplication and changes in environment. Combination, or interaction, of these two factors may certainly accelerate the fixation of functional divergence mutations in protein-coding genes. After gene duplication, two copies originate from an ancestral one, both holding identical nucleotide or amino acid compositions, and thus identical functions. Due to this redundancy, one gene copy can evolve free from selection, thereby exploring the phenotypic network neighboring ancestral genotypes (Ohno 1970), although see alternative, more sophisticated, models of evolution after gene duplication (Innan and Kondrashov 2010). Genes that present functional promiscuity, can duplicate and each copy specialize in one of a set of ancestral functions, given certain population genetics parameters, a process known as sub-functionalization (Ohno 1970; Lynch and Conery 2000; Lynch and Katju 2004; Conant and Wolfe 2006, 2008; Innan and Kondrashov 2010).

Environmental change can also lead to functional divergence, by allowing the fixation of advantageous mutations in the current environment that were kept cryptic in a previous environment. This functional diversification can lead to an ecological diversification as well. A case in point is the genome wide functional divergence in prokaryotes (Toft et al. 2009; Caffrey et al. 2012), which are also extraordinarily rich in biological diversity in terms of the number of species (Dykhuizen 1998; Gans et al. 2005), habitat range they colonize (Pikuta et al. 2007), or breadth of energy sources they can exploit (Pace 1997).

Identification of functional divergence from genomic data relies on the assumption that changes in a protein's function may leave signatures in their encoding genes. Inferring functional divergence using sequences can then be achieved by identifying changes that may have important impact on the performance of a gene. Methods to identify functional divergence rely precisely on this simple assumption. In the next sections we will discuss such methods, their application, and present illustrative examples.

5.2 Molecular Adaptation and Shifts in Selective Constraints

Identifying the signatures of adaptation at the molecular level is a challenging problem in biology. Most mutations in protein-coding genes, and indeed in regulatory regions, are fixed because they are neutral with regards to their effects on the organism's fitness (Kimura 1983). When the effective population sizes are finite, slightly deleterious and advantageous mutations can also become fixed in a neutral fashion (Ohta 1992). Because the fixation of mutations with strong effects on fitness is rare, it is paramount to develop models that can identify the few adaptive changes from an overwhelming majority of neutral mutations (Golding and Dean 1998). For this purpose, a plethora of methods to identify and quantify the strength of natural selection under explicit models of adaptive, purifying, and neutral evolution have been developed. However, such models remain very simplistic in inferring the constraints operating at the molecular level. There are several reasons that halt the identification of adaptive processes: (i) the relative contribution of molecules to the fitness of an organism is qualitatively, but not quantitatively, determined by these methods, (ii) the functional density of proteins, that is the functional contribution of amino acids to protein's function, is unknown, (iii) amino acids within proteins exercise reciprocal natural selection on one another in a highly tangled manner—a process termed molecular co-evolution, a process that is often neglected in most methods to identify selection, and (iv) selection constraints on one amino acid site may change throughout the evolution of a protein and in a continuous manner, a dynamic yet to be taken into account in most, if not all models of evolution.

Of the four limitations mentioned above, the change of the selective constraints on amino acid sites is conceptually crucial because methods usually assume that constraints on sites remain conserved throughout evolution. There are, however, a number of evolutionary scenarios under which selective constraints operating at particular sites of a protein may undergo dramatic turns. For example, functional changes in proteins are often reflected in changes on the tertiary protein structure. When a protein structure changes, so do the constraints on the different sites, such as, a site that was previously evolutionarily conserved in a group of organisms can become highly variable in another group (Figure 5.1) (Pupko and Galtier 2002). Based on this principle, shifts in selective constraints should also be apparent as changes in the rate of amino acid sites evolution. Therefore, testing for significant variations in the rates of evolution can uncover amino acid sites having undergone shifts in selective constraints and that may be key in the divergence of the protein from its ancestral function.

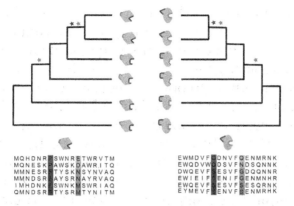

Figure 5.1. Changes in the selective constraints on an amino acid site of a protein can result from coadaptation dynamics among interacting amino acid sites. This figure represents two interacting proteins using cartoons to facilitate visualization (oval and hexagon shapes, respectively). Each of the proteins has a phylogenetic history represented by their corresponding topologies, as shown. In each of the protein cartoons we represent two regions, one labeled in green and another in red. The green regions coevolve historically between the proteins—they present the same phylogenetic pattern and therefore the same evolutionary variation profile. These two sites, however, do not always present compatible geometries—as they do not form complementary surfaces. The red-labeled regions are involved in physical interaction between both proteins and are the ones coevolving functionally (they geometrically complement one another across the phylogeny). For the sake of simplicity we represent each of the coevolving regions with an amino acid site in a multiple sequence alignment for each protein; letters represent one-letter amino acid code and they increase in size as they approach the site involved in a coevolutionary relationship. Green columns correspond to the green region in the protein cartoons and red columns to the red region in the protein cartoons. It is noticeable that important changes in amino acid properties in the red column of one protein are always corresponded by important ones in the red column of the other protein, while this is not the case for the green columns. The phylogenetic nodes at which changes in the red or green protein regions are indicated with stars, are color-coded according to the changing protein surface region. This figure has been adapted from a previous study (Fares et al. 2011).

Color image of this figure appears in the color plate section at the end of the book.

We mentioned earlier that a protein could diverge from its ancestral function by gene duplication if this is accompanied by a change in the ancestral environmental conditions. Since gene duplication originates two identical gene copies (although copies are not always born equal for a number of reasons (Lynch and Katju 2004)), one of the copies becomes selectively relaxed, thereby exploring alternative functions (Ohno 1970). This relaxed selective dynamics would lead to meaningful changes in the amino acid composition profiles in which a site that was functionally important, and thus subjected to strong selective constraints, will become now less important and more variable (Figure 5.2). Once the second copy has reached a new phenotype (function), selective constrains on that second copy would become stronger, precluding further changes on the

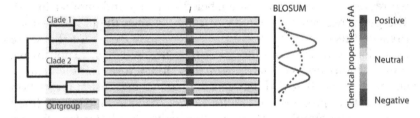

Figure 5.2. Shifts in the selective constraints on single amino acid sites. Proteins in which selective constraints change (or shift) show amino acid sites with significant evidence of shifts in the evolutionary rates in cluster 1 (cluster under study) compared to cluster 2 (background cluster). Cluster 1 and cluster 2 can either correspond to two sets of species postdating a gene duplication or two clades of orthologs. Identification of these selective shifts is possible through the comparison of each of the clusters to an outgroup. The change in the physico-chemical properties of amino acids from cluster 1 to cluster 2 is indicated by colored squares. If the transition scores are significantly more radical when comparing the outgroup to cluster 1 at that amino acid site of the protein, than when we compared the outgroup to cluster 2, then we consider the site to be under selection shifts.

Color image of this figure appears in the color plate section at the end of the book.

new set of functional sites. This process can lead to two alternative patterns: (a) the same set of sites have different, but equally important, functional roles in both of the copies resulting from gene duplication, leading to conserved but biochemically different amino acid compositions of that site at both gene copies; and (b) different set of sites would be functionally important in the two copies, leading to a pattern in which sites conserved in one copy are variable in the other and *vice versa*. The shift patterns of selective constraints can be metaphorically represented as the inverse of a fitness landscape. This landscape is populated by deep wells in which genotypes present new phenotypes. In the plane of the landscape, genotypes can cross from one point to another of the genotype space neutrally without accessing new phenotypes. Such genotypic travelling dynamics originate stochastic variations in the amino acid composition of amino acid sites. However, when a genotype falls in a functional well, the fitness distance to other neighboring genotypes increases and the transition from this new genotype to any of the others involves crossing a landscape region with a high fitness cost (Figure 5.3). The main difference between this trap landscape and a fitness landscape is that the former generates a population occupying a very robust position of the fitness landscape in which a substantial number of possible epistatic mutations is required to escape the trap, while in the latter, populations occupy a very volatile region of the fitness landscape in which a low number of mutations can push populations away from the adaptive peak. Eventually, a landscape populated with wells can become transformed into a conventional peaks-and-valleys landscape once the new phenotype has adapted to the environment.

A number of methods have been devised to identify functional divergence using the signatures of changes in the rates of evolution as proxy. The most widely used method

89

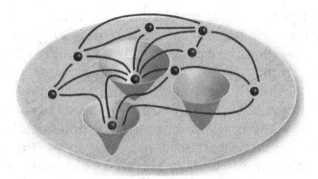

Figure 5.3. A trap landscape explains the emergence of novel phenotypes. In this landscape, transition between genotypes (symbolized here as solid black spheres) is neutral in terms of fitness cost. However, once a genotype reaches a phenotype which is not adapted to the environment (for example, a fitness well), this genotype becomes trapped in this phenotype and can only escape to another genotype provided enough number of mutations are produced. Eventually, this landscape can dynamically form a conventional peaks-and-valleys fitness landscape once the novel phenotype has adapted to the environment.

Color image of this figure appears in the color plate section at the end of the book.

is one that examines functional divergence after gene duplication. However, other methods have been designed to identify shifts in the selective constraints among organismal phylogenetic clades. Two such methods are described below with exemplifying case studies that illustrate their usefulness and applicability.

5.3 Identifying Functional Divergence Using Maximum Likelihood (ML)

Most methods devoted to the identification of functional divergence, and indeed the one described below, are based upon the assumption that the functional importance of sites is inversely proportional to the evolutionary rate of amino acid replacement (Graur and Li 2000). This is based on that functionally important sites are less amenable to fix amino acid replacements, as they drive proteins away from their ancestral well-adapted functions. Following this rationale, if a radical amino acid replacement takes place at conserved sites at some point of the evolution of a protein and becomes fixed thereafter, this change would be indicative of an adaptive process which may have involved a modification of the ancestral function. Changes in the selection regimes of particular sites of a protein are taken as evidence of functional divergence (Gu 1999, 2001; Knudsen and Miyamoto 2001; Susko et al. 2002).

Classically, important changes at particular regions of a protein are measured using the ratio between the rates of synonymous substitutions and that of non-synonymous replacements ($\omega = d_N/d_S$). When a protein is subject to Darwinian selection, d_N can exceed d_S, yielding vaues of $\omega > 1$. To identify statistically significant cases in which $\omega > 1$, this parameter was formally introduced as an explicit parameter of a Markov model of codon evolution (Goldman and Yang 1994; Muse and Gaut 1994). A number of models were developed that use such a parameter to identify signatures of Darwinian selection under discrete and continuous distributions (Yang and Nielsen 2002). Of these models, those concerning the estimation of ω in pre-specified branches of the tree and testing their differences with other branches are of particular interest to this chapter as they allow identifying selection shifts, and thus signatures of functional divergence (Yang and Nielsen 2002; Bielawski and Yang 2004; Zhang et al. 2005).

The use of ω to identify selection rests on two assumptions, namely that synonymous substitutions are neutral and that synonymous sites are not saturated, and thus the number of such substitutions is proportional to

91

the time of sequence divergence. However, recent studies are showing that synonymous substitutions can evolve under selective constraints, mainly dictated by the effect of such substitutions in translational efficiency or on the stability of the secondary structure of RNAs. The use of non-synonymous-to-synonymous rates ratio requires caution.

For long evolutionary distances, however, it becomes statistically convenient to use the patterns of amino acid replacements to identify selective changes among protein sequences. Amino acids provide a 20 letters code, corresponding to the 20 different amino acids, and do not rely on the assumption of neutrality, as is the case of synonymous sites. Methods that utilize amino acid replacements as indication of the rates of evolution compare the rates of amino acid substitutions among sites to test the heterogeneity of such rates along a protein sequence, a model formally represented by a gamma distribution (Uzzell and Corbin 1971; Gu et al. 1995; Yang 1996). In the gamma distribution there are two cases, those in which the rates heterogeneity among sites is conserved throughout the phylogeny (homogeneous gamma) and that in which this rate changes along the phylogeny (heterogeneous gamma). In the homogeneous gamma distribution, conserved sites and variable sites remain conserved and variable, respectively, across the entire phylogeny. In the heterogeneous gamma distribution, the rates of evolution of particular sites of the protein may change significantly in particular branches. It is the second of these gamma distributions, the heterogeneous gamma, the one on which models rest to identify Type I functional divergence, which is based on a Markov chain model (Gu 1999, 2001; Wang and Gu 2001). A difference in amino acid frequencies between sub-trees due to a change in both amino acid composition and the rate of amino acid evolution has also been interpreted as a signal of functional divergence, called Type II functional divergence, which is based on a Poisson model (Zhang et al. 2005; Gu et al. 2013). In this chapter, we will focus on Type I functional divergence.

5.3.1 Identifying Type I functional divergence

Type I functional divergence postulates that divergence in the functions between duplicate genes results in a shift in the selective constraints at some sites, which evolve at different rates in the lineage postdating gene duplication, but pre-dating speciation, compared to that within species clusters (Figure 5.2) (Gu 1999). The sites following these patterns are called F1 sites, while functional divergence unrelated sites are called F0 sites. In the two-states model developed by Gu (1999), the rates of F0 are similar for the two clusters postdating functional divergence ($R_1 = R_2$). Functional divergence-related sites (F1) show different rates for the two clusters resulting from gene duplication ($R_1 \neq R_2$). The key point to determine whether a gene has undergone functional divergence is calculating the coefficient of functional divergence (θ), which is the probability that a site belongs to F1 category of sites. This coefficient can then be used as an explicit parameter in which the likelihood of a model assuming $\theta > 0$ can be compared to that of a model assuming no functional divergence ($\theta = 0$). This comparison is done using the likelihood ratio test (LRT), which can be approached to a χ^2 distribution with 1 degree of freedom. Posterior Bayesian probabilities of a site belonging to F1 can be assigned to each site in the protein sequence, thereby building a site-specific profile based on posterior scores. This profile allows determining what sites have contributed to functional divergence providing that the difference in the log-likelihood values of the two hypotheses is significant.

The method of functional divergence is implemented in the program DIVERGE v1.4 and v2.0, which is a user-friendly package for the functional prediction of protein sequence data (http://www.xungulab.com). Another version (Diverge Version 3.0 (Gu et al. 2013)) has been recently released, although this remains under test. This program has been widely used (see for example, (Wang and Gu 2001; Gu et al. 2002; Gu and Vander Velden 2002; McNally and Fares 2007; Zheng et al. 2007; Zhou et al. 2007)) and runs under Linux and Windows operating systems. Of the many analyses conducted by Diverge, we exemplify the use of this program to identify signatures of shifts in the selective constraints at particular amino acid sites in the heat-shock protein GroEL after its duplication in bacteria (McNally and Fares 2007), which we have introduced in a previous chapter.

5.3.1.1 Functional divergence after GroEL duplication

In addition to the folding activity of GroEL, this protein has been reported to perform other unrelated folding functions, of which its immune eliciting function (Perschinka et al. 2003) is the one attracting most attention amongst microbiologists. The set of promiscuous functions of this protein is, however, substantially large, placing this protein at the focus of projects aiming at identifying proteins with moonlighting functions (see (Henderson et al. 2013) for a review of the subject). GroEL has been implicated in pathogenesis caused by infections with the bacterium *Chlamydiae trachomatis* (Karunakaran et al. 2003). In this bacterium, the gene encoding GroEL is duplicated two times, and thus we can identify three copies of this gene, named *groEL1*, *groEL2* and *groEL3* (Karunakaran et al. 2003). GroEL has been implicated in inflammatory processes caused by chlamydial infections (Peeling et al. 1998; Sanchez-Campillo et al. 1999; Lichtenwalner et al. 2004). Important differences between the *groEL* gene copies at the expression and functional levels have been reported: (i) there are differences in *groEL* copies expression profiles along the pathogenic cycle of *Chlamydiae*, and (ii) *groEL* gene copies exhibit different thermosensitivities.

Application of Diverge to identify functional divergence requires a minimum of four sequences from different related species or strains post-dating gene duplication—that is, each copy of *groEL* gene should be represented by at least four different strains or bacteria. Indeed, nine chlamydial species are available for *groEL* postdating its duplication (Figure 5.4) and the presence of all three copies in these nine species supports the duplication of *groEL* before the emergence of these species. The functional features of *groEL* as well as its evolutionary history makes this gene an ideal candidate in the testing of functional divergence.

To conduct functional divergence analyses in *groEL* from *Chlamydiae*, we performed the following steps:

1. We built a multiple sequence alignment including all three *groEL* gene copies using Clustal Omega (Sievers et al. 2011).
2. We stored the resulting alignment in "Fasta" format.
3. This alignment file was uploaded in Diverge by clicking on the option "Sequences".
4. Using the option "Clustering" and choosing Kimura 2-parameters model, we inferred a phylogeny (Figure 5.4). This phylogeny clearly showed each of the

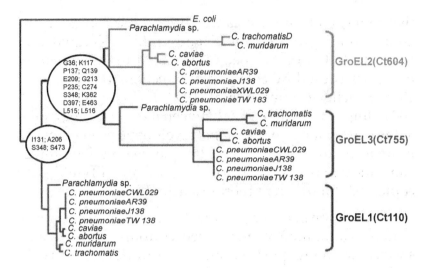

Figure 5.4. Analysis of selection shifts in GroEL from Chlamydial bacteria identifies sites under functional divergence. GroEL has duplicated twice in Chlamydia before the speciation of this bacterial strain. The three groups of GroEL paralogs are monophyletic. Diverge, a program to identify functional divergence after gene duplication, detected significant signatures of functional divergence after each duplication event. Amino acid sites exhibiting such divergence signatures are highlighted at the nodes after each gene duplication event.

Color image of this figure appears in the color plate section at the end of the book.

paralogs being monophyletic and the duplication events preceding speciation.

5. The option "Gu99" allowed us to estimate the parameter of functional divergence θ and the log-likelihood values for the hypotheses in support of functional divergence or otherwise.

6. Site-specific profiles can then be viewed by ticking the boxes corresponding to the different clusters compared.

7. Sites with higher posterior Bayesian probabilities, and thus candidates for functional divergence sites, can

then be displayed in the alignment and structure of the protein, when available.

Analysis of functional divergence after each duplication events provides evidence in support of a change in the selective constraints after gene duplication (Table 5.1). The parameter of functional divergence is significantly larger than 0 and the fit of the model of functional divergence to the data is significantly larger than a model that does not account for functional divergence (Table 5.1).

Table 5.1. Analysis of functional divergence after the two *groEL* gene duplication events. The model with functional divergence was tested against a model assuming no functional divergence using the likelihood ratio test (LRT). Twice the difference in the likelihood values between these two models was compared to a chi-square distribution with 1 degree of freedom. The parameters estimated in the models was that of functional divergence (Θ) and the heterogeneity in the distribution of substitution rates along the sequences (α).

Comparison	$\Theta \pm SE(\theta)$	α	LRT	P(LRT)
GroEL1 vs GroEL2-3	0.414 ± 0.117	3.448	12.484	$<< 0.001$
GroEL2 vs GroEL3	0.441 ± 0.073	3.105	36.014	$<< 0.001$

5.3.2 Identification of functional divergence after ecological diversification

The previous section was devoted to identifying functional divergence after gene duplication. However, other situations may also lead to shifts in the evolutionary rates of proteins. A chief factor is the change in the ecological conditions of organisms. Indeed, bacteria inhabit a bewildering range of ecological environments, each of which imposes specific selective constraints on a set of proteins. Major changes in the ecological conditions of organisms can lead to relaxed constraints on ancestral functions, although this process is poorly characterized. For example, endosymbiotic bacteria of insects have lost many of the genes their free-relatives needed to obtain nutrients from the environment (Moran 2002), but have also experienced shifts in the functions encoded by certain genes (Toft et al. 2009; Williams et al. 2010).

Identifying the fraction of genetic variation associated with functional changes remains a challenging problem. In the case of bacteria, whole-genome analysis must take into account the high likelihood of horizontal gene transfer (Ochman et al. 2000), which means that genes in bacteria often conflict in their phylogenetic profiles, and thus with the overall species tree (Dagan and Martin 2006). This problem may well halt analyses of functional divergence since such analyses heavily rely on a phylogeny to determine the branches upon which a trait arose.

To identify genome-wide functional divergence between orthologous groups of sequences using molecular phylogenetics, Caffrey and colleagues recently developed a simple method, which is implemented in the program CAFS version 1.0 (Clustering Analysis of Functional Shifts (Caffrey et al. 2012)). This method relies on the general assumption that functional sites are constrained by selection and should therefore evolve slowly. The method is distance-based and explores a bifurcating phylogenetic tree, testing for functional divergence at each node by comparing the two downstream clades to an outgroup in order to identify sites at which substitution rates per amino acid have shifted (Toft et al. 2009; Williams et al. 2010; Caffrey et al. 2012).

Caffrey et al. (2012) conducted a proteome-wide analysis of functional divergence by inferring patterns of radical amino acid changes for each protein individually, and then clustering species of bacteria according to the functional categories in which sufficient evidence of functional divergence arose (identified using Clusters of Orthologous Groups COG (Tatusov et al. 1997)). To do so, they used 750 bacterial proteomes. This set included bacteria from various ecological sources, providing thereby a good dataset for identifying ecological-related functional divergence.

The flow of functional divergence calculations using CAFS version 1.0 is shown in Figure 5.5. In brief, once sequence alignments are uploaded to the program, the user may decide whether to provide a phylogenetic tree or not. The user can provide the phylogenetic tree in Newick format. If the user decides not to provide a phylogeny, then this phylogenetic tree is inferred for each of the protein alignments. Trees are built under the JTT model of protein sequence evolution (Jones et al. 1992) and using the BIONJ tree building method (Gascuel 1997). For each alignment tested, ancestral sequences (sequences at the internal nodes of the phylogeny) are reconstructed using maximum likelihood under the JTT model, as implemented with the use of BIO++ libraries (Dutheil et al. 2006). To identify functional divergence, the algorithm walks on the tree and calculates functional divergence scores at each node of the tree. Clades on either side of the bifurcation are compared to the closest available outgroup with respect to an amino acid substitution matrix. Scores of functional divergence are given by:

$$FD_{score} = \frac{X_1 - X_2}{s_{X_1 - X_2}} \tag{5.1}$$

Where $X_{1,2}$ are the mean substitution scores for mutations from clades on either side of the bifurcation in the phylogenetic tree relative the outgroup and the standard

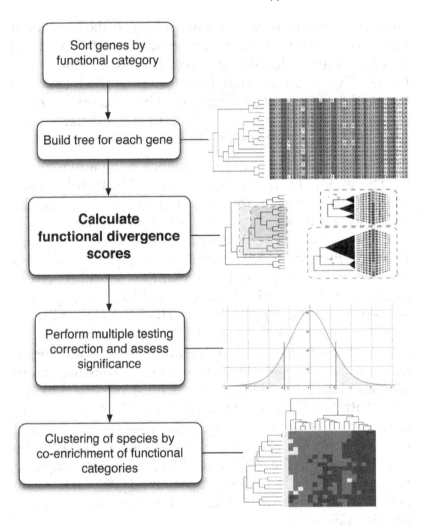

Figure 5.5. Flow chart shows the information input and output from the program CAFS (Available at hhtp//bioinf.gen.tcd.ie/~faresm), a software that identifies signatures of functional divergence in groups of ortologous sequences and duplicates.

Color image of this figure appears in the color plate section at the end of the book.

error for unequal sample sizes with unequal variances are given by:

$$S_{X_1-X_2} = \sqrt{\frac{S_1^2}{n_1} + \frac{S_2^2}{n_2}} \tag{2}$$

To test the significance of the FD scores, simulated sequence alignments are created using the JTT model according to the gene-specific phylogenetic tree calculated above. Simulated alignments are tested according to equation (1), resulting in a null distribution of the test score against which P-values for the real data can be evaluated, with significance taken at the 5% level. These values are then corrected for multiple testing by the False Discovery Rate method. Following this procedure, branches on the tree that still possess at least one significant site are considered to be under functional divergence.

Once all alignments have been analyzed, three different enrichment tests are performed: for each species, for each functional category, and for species-by-functional category. These tests are performed under a chi-square distribution:

$$\chi^2 = \frac{\sum_{i=1}^{n}(O_i-E_i)^2}{E_i} \tag{3}$$

where O_i is the observed frequency of genes under functional divergence, E_i is the expected frequency and n is the number of possible outcomes for each event. The result is three tests that assess:

1. Whether each species is enriched (either under or over enrichment) relative to all other species: that is, whether some species experience more or less functional divergence than others.

2. Whether each functional category is enriched (either under or over enrichment) relative to all other categories: whether some groups of genes undergo more or less functional divergence than others.
3. Whether each functional category in each species is enriched (either under or over enrichment) relative to all other categories in each species: whether particular species undergo more or less functional divergence in particular categories than others.

5.3.2.1 Interaction with a host constrains functional change in pathogenic bacteria

One of the main questions that identifying shifts in the rates of evolution may help to answer is that concerning the fingerprints of ecological adaptations in proteins functions. Functional divergence can be run on a folder containing the multiple sequence alignments for the entire Proteome. The Method to detect functional divergence is implemented in the program CAFS v1.0. This program can be downloaded from (http://bioinf.gen.tcd.ie/~faresm/software). The program contains the binaries for Mac 10.6 or higher operating system. Running this program for the entire proteome requires alignments to be placed in a single folder. Once a terminal is open, CAFS can be run typing:

" ./cafs –F my_folder –t my_tree –c 0.001 –m BLOSUM80.tab"

my_folder refers to the folder containing the multiple sequences alignments for all orthologous for each protein in the proteome. User could provide an initial tree if the species tree is known, otherwise the program infers a maximum likelihood phylogenetic tree for each gene.

Inferring Functional Divergence in Protein Sequences

The threshold probability for the functional divergence coefficient is given by the parameter –c. Different models of amino acid evolution can be pre-specified by the user using the parameter –m.

Once we run CAFS on the full genome of *E. coli* and 750 orthologous gamma proteobacteria, a functional divergence score is obtained for each of the alignments. To calculate the enrichment status of each species for functional divergence, we calculate a null background proportion of successful tests of functional divergence for each individual species using chi-squared tests. Then, chi-square is also used to identify meaningful associations between the enrichment patterns of functional divergence (enriched, impoverished or neither) and organism lifestyle, resulting in the information provided in Table 5.2. One of the most interesting results from Table 5.2 is the impoverishment of host-interacting bacteria for functional divergence compared to their free-living cousins. This result can be explained in terms of the more specialized ecological nature of the host intra-cellular milieu in which these bacteria live. In particular, symbiotic bacteria have lost many of the genes encoding environment-exploitation specialized proteins, as they no longer require exploiting as a wide a range of nutrients source as free-living bacteria (Moran 2002).

Analysis of specific functional categories allows identifying genes with the greatest consistent difference in functional divergence between host-associated bacteria and free-living bacteria. Interestingly, comparison of these two sets of bacteria shows that genes involved in vesicular transport and secretion systems (category U) are enriched for functional divergence in host-associated bacteria but neither enriched or impoverished in free-living bacteria, while signal transduction genes are impoverished in host-associated bacteria but enriched in their free-living relatives.

Table 5.2. Enrichment analyses for functional divergence in bacteria. We studied the correlation between the lifestyle of a number of bacterial clusters (e.g., mesophiles, pathogenic, mutualistic or non-pathogenic) with their profiles of functional divergence enrichment. Significant enrichment or impoverishment (−) of a clade of bacteria for functional divergence is indicated with stars (*). * indicates $P < 0.05$, and *** indicates $P < 0.0001$.

Lifestyle	Comparison	Enriched	Impoverished	Neither	Significance
Psychrophile	Mesophile	2/61	1/66	6/433	N.S.
Thermophile	Mesophile	7/61	1/66	22/433	N.S.
Pathogen	Non-pathogen	22/77	47/38	272/294	***(−)
Intracellular Pathogen	Other pathogen	0/22	4/43	26/246	N.S.
Symbiont	Non-symbiont	4/95	13/72	36/530	*
*Intracellular symbiont	All others	4/224	15/133	14/360	***(−)
All interactors	Free-living	44/55	70/15	410/156	***(−)

These patterns correlate very well with the life style of these bacteria, as pathogens use elaborate secretion systems for delivering toxins and virulence factors to the environment provided by their host (Baron 2010). Conversely, symbionts supplement their host's diet with essential nitrogen compounds as part of their beneficial symbiotic association (Douglas 1998; Sandstrom et al. 2000).

5.3.2.2 From Systems to Molecules

To gain more insights into the role of functional divergence in the ecological diversification of bacteria, one could cluster species using two-dimensional hierarchical clustering according to the enrichment status associated with each species and functional category. The resulting heat-map (Figure 5.6a) provides an intuitive way of interpreting the results at the general level while identifying particularities that do not follow general trends. For example, Figure 5.6 illustrates an extreme conservation among informational genes, in particular those involved in biogenesis (category J), while it highlights lineage-specific exceptions to the general trends, such as those provided by Bartonella.

The genus of Bartonella (Figure 5.6b) defines a group of intracellular parasites that infect and replicate in erythrocytes (Anderson and Neuman 1997). Of the Bartonella species used in this study (Caffrey et al. 2012), only one, *B. baciliformis*, is enriched for functional divergence in cell motility genes (N), with the other species being impoverished (2 species) or neither enriched or impoverished (1 species). Interestingly, *B. baciliformis* is the only member of the Bartonella family that posseses flagella (Brenner et al. 1991). It noteworthy that erythrocytes lack an active cytoskeleton, meaning that they cannot take bacteria from the environment by invagination

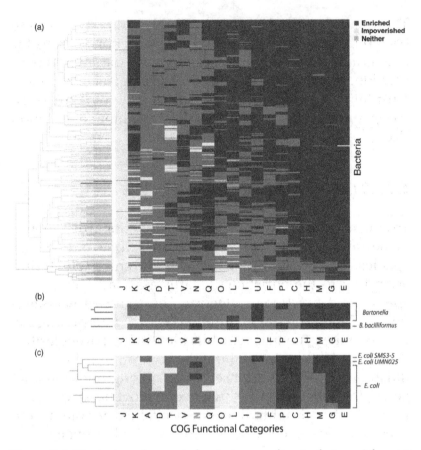

Figure 5.6. Clustering of genes in bacteria according to their enrichment profiles for functional divergence. Each of the columns in the heatmap indicates a functional category according to the Cluster of Orthologous Groups. Each row represents one bacterial strain. Blue, yellow and grey represent functional categories enriched for functional divergence, impoverished and neutral, respectively. (a) Analysis of functional divergence for 750 bacterial genomes. (b) Analysis of functional divergence enrichement for the cluster of *Bartonella* pathogenic bacteria. (c) Identification of functional divergence among closely related *Escherichia coli* strains.

Color image of this figure appears in the color plate section at the end of the book.

106

(Dramsi and Cossart 1998), and thus infection of erythrocyte by Bartonella is an active process (Dehio 2001). The use of flagella by *B. baciliformis* to infect erythrocytes (Scherer et al. 1993) makes it more efficient than other species, infecting up to 80% of erythrocytes (Scherer et al. 1993; Ihler 1996; Dehio 2001). Therefore, the enrichment in motility for *B. baciliformis* is a clear case of functional divergence to adapt to the environment and attests to the practical usefulness of genome-wide identification of patterns of selection shifts.

References

Anderson, B.E. and M.A. Neuman. 1997. *Bartonella* spp. as emerging human pathogens. Clinical Microbiology Reviews 10: 203–219.

Baron, C. 2010. Antivirulence drugs to target bacterial secretion systems. Current Opinion in Microbiology 13: 100–105. doi: 10.1016/j.mib.2009.12.003.

Bielawski, J.P. and Z. Yang. 2004. A maximum likelihood method for detecting functional divergence at individual codon sites, with application to gene family evolution. J Mol Evol 59: 121–132. doi: 10.1007/s00239-004-2597-8.

Brenner, D.J., S.P. O'Connor, D.G. Hollis, R.E. Weaver and A.G. Steigerwalt. 1991. Molecular characterization and proposal of a neotype strain for Bartonella bacilliformis. Journal of Clinical Microbiology 29: 1299–1302.

Caffrey, B., T.A. Williams, X. Jiang, C. Toft, K. Hokamp and M.A. Fares. 2012. Proteome-wide analysis of functional divergence in bacteria: Exploring a host of ecological adaptations. PLoS One Submitted.

Conant, G.C. and K.H. Wolfe. 2006. Functional partitioning of yeast co-expression networks after genome duplication. PLoS Biol 4: e109. doi: 05-PLBI-RA-1196R3 [pii].

Conant, G.C. and K.H. Wolfe. 2008. Turning a hobby into a job: how duplicated genes find new functions. Nat Rev Genet 9: 938–950. doi: 10.1038/nrg2482 [pii].

Dagan, T. and W. Martin. 2006. The tree of one percent. Genome Biol 7: 118. doi: 10.1186/gb-2006-7-10-118.

Dehio, C. 2001. Bartonella interactions with endothelial cells and erythrocytes. Trends in Microbiology 9: 279–285.

Douglas, A.E. 1998. Nutritional interactions in insect-microbial symbioses: aphids and their symbiotic bacteria Buchnera. Annual Review of Entomology 43: 17–37. doi: 10.1146/annurev.ento.43.1.17.

Dramsi, S. and P. Cossart. 1998. Intracellular pathogens and the actin cytoskeleton. Annual Review of Cell and Developmental Biology 14: 137–166. doi: 10.1146/annurev.cellbio.14.1.137.

Dutheil, J., S. Gaillard, E. Bazin, S. Glemin, V. Ranwez, N. Galtier and K. Belkhir. 2006. Bio++: a set of C++ libraries for sequence analysis, phylogenetics, molecular evolution and population genetics. BMC Bioinformatics 7: 188. doi: 10.1186/1471-2105-7-188.

Dykhuizen, D.E. 1998. Santa Rosalia revisited: why are there so many species of bacteria? Antonie van Leeuwenhoek 73: 25–33.

Fares, M.A., M.X. Ruiz-Gonzalez and J.P. Labrador. 2011. Protein coadaptation and the design of novel approaches to identify protein-protein interactions. IUBMB Life 63: 264–271. doi: 10.1002/iub.455.

Gans, J., M. Wolinsky and J. Dunbar. 2005. Computational improvements reveal great bacterial diversity and high metal toxicity in soil. Science 309: 1387–1390. doi: 10.1126/science.1112665.

Gascuel, O. 1997. BIONJ: an improved version of the NJ algorithm based on a simple model of sequence data. Mol Biol Evol 14: 685–695.

Golding, G.B. and A.M. Dean. 1998. The structural basis of molecular adaptation. Mol Biol Evol 15: 355–369.

Goldman, N. and Z. Yang. 1994. A codon-based model of nucleotide substitution for protein-coding DNA sequences. Mol Biol Evol 11: 725–736.

Graur, D. and W.H. Li. 2000. Fundamentals of Molecular Evolution. Sunderland, Massachusetts: Sinauer Associates, Inc.

Gu, J., Y. Wang and X. Gu. 2002. Evolutionary analysis for functional divergence of Jak protein kinase domains and tissue-specific genes. J Mol Evol 54: 725–733. doi: 10.1007/s00239-001-0072-3.

Gu, X. 1999. Statistical methods for testing functional divergence after gene duplication. Mol Biol Evol 16: 1664–1674.

Gu, X. 2001. Maximum-likelihood approach for gene family evolution under functional divergence. Mol Biol Evol 18: 453–464.

Gu, X., Y.X. Fu and W.H. Li. 1995. Maximum likelihood estimation of the heterogeneity of substitution rate among nucleotide sites. Mol Biol Evol 12: 546–557.

Gu, X. and K. Vander Velden. 2002. DIVERGE: phylogeny-based analysis for functional-structural divergence of a protein family. Bioinformatics 18: 500–501.

Gu, X., Y. Zou, Z. Su, W. Huang, Z. Zhou, Z. Arendsee and Y. Zeng. 2013. An update of DIVERGE software for functional divergence analysis of protein family. Mol Biol Evol 30: 1713–1719. doi: 10.1093/molbev/mst069.

Henderson, B., M.A. Fares and P.A. Lund. 2013. Chaperonin 60: a paradoxical, evolutionarily conserved protein family with multiple moonlighting functions. Biological Reviews of the Cambridge Philosophical Society 88: 955–987. doi: 10.1111/brv.12037.

Ihler, G.M. 1996. Bartonella bacilliformis: dangerous pathogen slowly emerging from deep background. FEMS Microbiology Letters 144: 1–11.

Innan, H. and F. Kondrashov. 2010. The evolution of gene duplications: classifying and distinguishing between models. Nat Rev Genet 11: 97–108. doi: 10.1038/nrg2689 [pii].

Jones, D.T., W.R. Taylor and J.M. Thornton. 1992. The rapid generation of mutation data matrices from protein sequences. Computer Applications in the Biosciences: CABIOS 8: 8.

Karunakaran, K.P., Y. Noguchi, T.D. Read, A. Cherkasov, J. Kwee, C. Shen, C.C. Nelson and R.C. Brunham. 2003. Molecular analysis of the multiple GroEL proteins of Chlamydiae. Journal of Bacteriology 185: 1958–1966.

Kimura, M. 1983. The neutral theory of molecular evolution. Cambridge University Press, London.

Knudsen, B. and M.M. Miyamoto. 2001. A likelihood ratio test for evolutionary rate shifts and functional divergence among proteins. Proc Natl Acad Sci U S A 98: 14512–14517. doi: 10.1073/pnas.251526398.

Lichtenwalner, A.B., D.L. Patton, W.C. Van Voorhis, Y.T. Sweeney and C.C. Kuo. 2004. Heat shock protein 60 is the major antigen which stimulates delayed-type hypersensitivity reaction in the macaque model of Chlamydia trachomatis salpingitis. Infection and Immunity 72: 1159–1161.

Lynch, M. and J.S. Conery. 2000. The evolutionary fate and consequences of duplicate genes. Science 290: 1151–1155. doi: 8976 [pii].

Lynch, M. and V. Katju. 2004. The altered evolutionary trajectories of gene duplicates. Trends Genet 20: 544–549. doi: S0168-9525(04)00250-1 [pii] 10.1016/j.tig.2004.09.001.

McNally, D. and M.A. Fares. 2007. *In silico* identification of functional divergence between the multiple groEL gene paralogs in Chlamydiae. BMC Evol Biol 7: 81. doi: 10.1186/1471-2148-7-81.

Moran, N.A. 2002. Microbial minimalism: genome reduction in bacterial pathogens. Cell 108: 583–586.

Muse, S.V. and B.S. Gaut. 1994. A likelihood approach for comparing synonymous and nonsynonymous nucleotide substitution rates, with application to the chloroplast genome. Mol Biol Evol 11: 715–724.

Ochman, H., J.G. Lawrence and E.A. Groisman. 2000. Lateral gene transfer and the nature of bacterial innovation. Nature 405: 299–304. doi: 10.1038/35012500.

Ohno, S. 1970. Evolution by Gene duplication. Berlin: Springer-Verlag.

Ohta, T. 1992. The nearly neutral theory of molecular evolution. Annual Review of Ecology and Systematics 23: 24.

Pace, N.R. 1997. A molecular view of microbial diversity and the biosphere. Science 276: 734–740.

Peeling, R.W., R.L. Bailey, D.J. Conway, M.J. Holland, A.E. Campbell, O. Jallow, H.C. Whittle and D.C. Mabey. 1998. Antibody response to the 60-kDa chlamydial heat-shock protein is associated with scarring trachoma. The Journal of Infectious Diseases 177: 256–259.

Perschinka, H., M. Mayr, G. Millonig, C. Mayerl, R. van der Zee, S.G. Morrison, R.P. Morrison, Q. Xu and G. Wick. 2003. Cross-reactive B-cell epitopes of microbial and human heat shock protein 60/65 in atherosclerosis. Arteriosclerosis, thrombosis, and vascular biology 23: 1060–1065. doi: 10.1161/01.ATV.0000071701.62486.49.

Pikuta, E.V., R.B. Hoover and J. Tang. 2007. Microbial extremophiles at the limits of life. Critical Reviews in Microbiology 33: 183–209. doi: 10.1080/10408410701451948.

Pupko, T. and N. Galtier. 2002. A covarion-based method for detecting molecular adaptation: application to the evolution of primate mitochondrial genomes. Proc Biol Sci 269: 1313–1316. doi: 10.1098/rspb.2002.2025.

Sanchez-Campillo, M., L. Bini, M. Comanducci, R. Raggiaschi, B. Marzocchi, V. Pallini and G. Ratti. 1999. Identification of immunoreactive proteins of Chlamydia trachomatis by Western blot analysis of a two-dimensional electrophoresis map with patient sera. Electrophoresis 20: 2269–2279. doi: 10.1002/(SICI)1522-2683(19990801)20:11<2269::AID-ELPS2269>3.0.CO;2-D.

Sandstrom, J., A. Telang and N.A. Moran. 2000. Nutritional enhancement of host plants by aphids—a comparison of three aphid species on grasses. Journal of Insect Physiology 46: 33–40.

Scherer, D.C., I. DeBuron-Connors and M.F. Minnick. 1993. Characterization of Bartonella bacilliformis flagella and effect of antiflagellin antibodies on invasion of human erythrocytes. Infection and Immunity 61: 4962–4971.

Sievers, F., A. Wilm, D. Dineen, et al. 2011. Fast, scalable generation of high-quality protein multiple sequence alignments using Clustal Omega. Molecular Systems Biology 7: 539. doi: 10.1038/msb.2011.75.

Susko, E., Y. Inagaki, C. Field, M.E. Holder and A.J. Roger. 2002. Testing for differences in rates-across-sites distributions in phylogenetic subtrees. Mol Biol Evol 19: 1514–1523.

Tatusov, R.L., E.V. Koonin and D.J. Lipman. 1997. A genomic perspective on protein families. Science 278: 631–637.

Toft, C., T.A. Williams and M.A. Fares. 2009. Genome-wide functional divergence after the symbiosis of proteobacteria with insects unraveled through a novel computational approach. PLoS Comput Biol 5: e1000344. doi: 10.1371/journal.pcbi.1000344.

Uzzell, T. and K.W. Corbin. 1971. Fitting discrete probability distributions to evolutionary events. Science 172: 1089–1096.

Wang, Y. and X. Gu. 2001. Functional divergence in the caspase gene family and altered functional constraints: statistical analysis and prediction. Genetics 158: 1311–1320.

Williams, T.A., F.M. Codoner, C. Toft and M.A. Fares. 2010. Two chaperonin systems in bacterial genomes with distinct ecological roles. Trends Genet 26: 47–51. doi: 10.1016/j.tig.2009.11.009 S0168-9525(09)00248-0 [pii].

Yang, Z. 1996. Statistical properties of a DNA sample under the finite-sites model. Genetics 144: 1941–1950.

Yang, Z. and R. Nielsen. 2002. Codon-substitution models for detecting molecular adaptation at individual sites along specific lineages. Mol Biol Evol 19: 908–917.

Zhang, J., R. Nielsen and Z. Yang. 2005. Evaluation of an improved branch-site likelihood method for detecting positive selection at the molecular level. Mol Biol Evol 22: 2472–2479. doi: 10.1093/molbev/msi237 [pii].

Zheng, Y., D. Xu and X. Gu. 2007. Functional divergence after gene duplication and sequence-structure relationship: a case study of G-protein alpha subunits. Journal of Experimental Zoology Part B, Molecular and Developmental Evolution 308: 85–96. doi: 10.1002/jez.b.21140.

Zhou, H., J. Gu, S.J. Lamont and X. Gu. 2007. Evolutionary analysis for functional divergence of the toll-like receptor gene family and altered functional constraints. J Mol Evol 65: 119–123. doi: 10.1007/s00239-005-0008-4.

Chapter 6

The Influence of Recombination on the Estimation of Selection from Coding Sequence Alignments

Miguel Arenas[1,*] and *David Posada*[2]

The nonsynonymous synonymous substitution rate ratio (dN/dS) is commonly used to detect selection in coding sequences at both global (entire genes) and local (codon position) levels. Among a large variety of organisms, this estimation is frequently performed from data sets where recombination has likely occurred. However, several studies

[1] Centre for Molecular Biology "Severo Ochoa", Consejo Superior de Investigaciones Científicas (CSIC). C/Nicolás Cabrera, 1. Cantoblanco. E-28049. Madrid, Spain.
Email: marenas@cbm.uam.es
[2] Department of Biochemistry, Genetics and Immunology, University of Vigo. 36310 Vigo, Spain.
Email: dposada@uvigo.es
* Corresponding author

have shown that recombination may bias selection estimates at local level, in particular by increasing the number of false positively selected sites. Here we describe this bias and provide alternative strategies to estimate dN/dS in the presence of recombination. Several practical examples are also included to illustrate this procedure.

6.1 Introduction

Signatures of global (entire gene) and local (position-specific) selection can be detected in coding data by using the nonsynonymous synonymous substitution rate ratio (hereafter, dN/dS) (Nei and Gojobori 1986). Codons with $dN/dS = 1$, <1 or >1 are usually assumed to be evolving under neutral evolution, negative selection or positive selection, respectively. While the global dN/dS may provide a general measure of molecular adaptation in the entire alignment, local dN/dS is more appropriate to detect selection at particular sites. For example, specific codon sites that encode amino acid residues closely involved in the protein activity, usually evolve under negative selection in order to keep protein function (Kim et al. 2007; Chen and Sun 2011).

Estimation of dN/dS is common in evolutionary analyses of many organisms, including highly recombinant viruses and bacteria (Kryazhimskiy et al. 2008; Perez-Losada et al. 2009; Bozek and Lengauer 2010; Castillo-Ramirez et al. 2011; Chen and Sun 2011; Perez-Losada et al. 2011). Nevertheless, several studies have unveiled the effect of recombination on the estimation of dN/dS (Anisimova et al. 2003; Shriner et al. 2003; Kosakovsky Pond et al. 2006b; Arenas and Posada 2010). This influence was explained as a consequence of ignoring recombination in traditional phylogenetic analysis

(Schierup and Hein 2000b; Schierup and Hein 2000a; Posada and Crandall 2002). In particular, Schierup and Hein have showed that recombination biases the inferred phylogenetic trees resulting in the spurious inference of larger terminal branch lengths, smaller times to the most recent common ancestor (MRCA), loss of molecular clock and overestimation of the substitution rate heterogeneity. On the other hand, Posada and Crandall (2002) showed that recombination may result in the inference of wrong tree topologies, which in turn might affect the estimation of *dN/dS*.

In this chapter, we will summarize the influence of recombination on the inference of selection, and the potential effect of intracodon recombination. We will also describe an alternative strategy to infer *dN/dS* values where recombination is considered and show several practical examples.

6.2 Influence of Recombination on the Estimation of Global Selection

Several studies have described the effect of recombination on the global *dN/dS* estimation (Anisimova et al. 2003; Shriner et al. 2003; Arenas and Posada 2010) from sequence alignments. In general, in these studies coding sequences are simulated under different levels of *dN/dS* and recombination rates. Then, global *dN/dS* values are estimated using different methods, like codon models (Yang 1999) or pairwise comparisons (Nei and Gojobori 1986). Overall, these studies indicate that recombination does not bias the estimation of the global *dN/dS*. Moreover, our own simulations indicate that this is also the estimation of branch-specific *dN/dS* values (unpublished results).

114

6.3 Influence of Recombination on the Estimation of Local Selection

Signatures of selection may change across codon sites and several methods have been developed for the inference of site-specific or local *dN/dS* values (Yang 1999; Kosakovsky Pond and Frost 2005a). Logically, single sites provide less information than entire sequences and consequently, the estimation of local *dN/dS* may be less robust, especially if the sequences are similar.

The effect of recombination on local *dN/dS* has also been studied (Anisimova et al. 2003; Shriner et al. 2003; Kosakovsky Pond et al. 2006b; Arenas and Posada 2010). Again, these authors simulated coding data under a variety of levels of recombination (but assuming that recombination only occurs between codons; see next section) and variable *dN/dS* across sites based on heterogeneous codon substitution models (Yang et al. 2000; Yang and Nielsen 2002). Then, they used likelihood ratio tests (LRTs) to compare neutral and selective models and tried to identify positively-selected sites (PSS). They observed that the LRTs were robust to low levels of recombination, but favored the spurious inference of selection when recombination was large. In particular, the number of false PSS increased as a function of the amount of recombination simulated (Anisimova et al. 2003; Shriner et al. 2003; Kosakovsky Pond et al. 2006b; Arenas and Posada 2010; Castillo-Ramirez et al. 2011). Indeed, recombination may result in incorrect tree heights and when *dS* is held constant (as it is in most popular codon models), this variation in tree length can be interpreted as variation in *dN*, and consequently, in variation of the *dN/dS* ratios (Anisimova et al. 2003). However, a "two-rate" or "dual" model, which does not hold *dS* constant, showed a much better behaviour, with more power to detect selection and less false PSS (Arenas and Posada 2010). The use of "dual" models, implemented for example in the *Hyphy* package (Kosakovsky Pond et al. 2005), is therefore recommended to estimate local *dN/dS* when recombination is present. Alternatively, one could try to consider the different recombination fragments separately (see section 5).

6.4 Influence of Intracodon Recombination Breakpoints on the Estimation of Selection

If recombination breakpoints along coding regions were uniformly distributed, we would expect 2/3 of all recombination events occurring within codons. While recombination breakpoints are not uniformly distributed across the genome (Wiuf and Posada 2003), we still expect them to occur frequently inside codons. In fact, recently Behura and Severson (2013) provided evidence for intracodon recombination in Dengue virus (DENV). Depending on the codons involved, some of these intracodon recombination events will result in nonsynonymous changes (see Figure 6.1), which could potentially bias the estimation of *dN/dS* ratios. Extensive simulations performed with the coalescent-based program *NetRecodon* showed that intracodon recombination *per se* does not make a strong effect on the estimation of global or local *dN/dS* at typical levels of nucleotide diversity (Arenas and Posada 2010).

Of course, intracodon recombination may generate nonsynonymous changes but the number of these changes is low with respect to the number of changes introduced by the substitution process. In addition, intracodon recombination events that occur between similar sequences will not have much effect, as most will occur between identical codons. Not surprisingly, intracodon recombination has

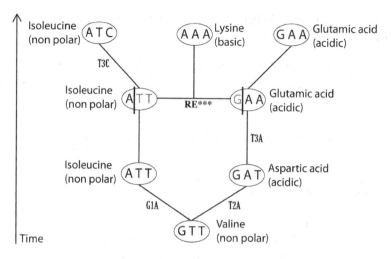

Figure 6.1. Example of how intracodon recombination can result in a nonsynonymous change. The scheme represents the arbitrary evolutionary history of a single codon. The codon GTT at the root of the tree evolves along the different branches according to some substitution processes. A recombination event (RE***) is introduced at a given point in time between the first and the second codon positions. Note that the physicochemical properties of the amino acid encoded by the recombinant codon, clearly differs in this case from the properties of the amino acids encoded by the parental codons.

been suggested to play a significant role in maintaining extensive purifying selection in Dengue viruses (Behura and Severson 2013).

6.5 Estimation of *dN/dS* from Recombining Sequences

We have shown that recombination can inflate the apparent number of codons under positive selection. Although such biases could be ameliorated by using "dual" models, a more efficient strategy consists in the explicit consideration of recombination. We have proposed a particular strategy for the latter (Figure 6.2):

1) Detection of recombination breakpoints, using *GARD* (Kosakovsky Pond et al. 2006a) (implemented in

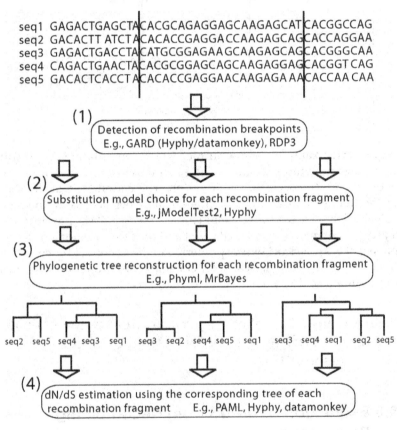

Figure 6.2. Estimation of *dN/dS* accounting for recombination from a coding alignment. (1) Given a coding alignment, recombination breakpoints can be detected using *GARD* (*Hyphy/Datamonkey* web server) or *RDP3*. In the example, breakpoints were detected at positions 12 and 33. (2) A substitution model should be selected for each recombinant fragment. (3) Given the substitution model, a phylogenetic tree can be reconstructed for each fragment. (4) The *dN/dS* estimation can be now performed by using the corresponding phylogenetic tree.

Hyphy and the *Datamonkey* web server (http://www. datamonkey.org/) (Kosakovsky Pond and Frost 2005b)) or *RDP3* (Martin et al. 2010) for additional recombination breakpoints detection tools see Martin et al. (2011).

2) The original alignment is split into partial alignments, one for each recombinant fragment, according to the detected recombination breakpoints.

3) A phylogenetic tree can be reconstructed for each partial alignment according to the corresponding substitution model.

4) *dN/dS* can be estimated specifying the corresponding phylogenetic tree for each fragment (e.g., using *PAML* or *Hyphy/Datamonkey* (Kosakovsky Pond and Frost 2005a)).

Actually, the *Datamonkey* web server and the *Hyphy* package have automated this whole procedure described above. For example, we have applied this methodology to estimated molecular adaptation (both global and local *dN/dS*) accounting for recombination in HIV-1 gp120 datasets collected from different ethnic subgroups to conclude intrinsic differences among races in immune response (Perez-Losada et al. 2009); and from datasets collected from a Phase III AIDS vaccine trial in Bangkok to analyze phylodynamics of the HIV-1 in Thailand (Perez-Losada et al. 2011).

An alternative methodology to estimate *dN/dS* accounting for recombination is the approximate Bayesian computation (ABC) approach (see for a review, Beaumont 2010). ABC allows the estimation of multiple evolutionary parameters accounting for recombination. This approach is based on extensive computer simulations and provides an alternative for those analyses where the likelihood function cannot be computed. Simulations can be carried out according to a prior distribution for dN/dS (among other priors for other parameters) and then, by a regression or a rejection method, a posterior distribution can be computed to obtain the parameter estimates (Beaumont et al. 2002; Beaumont 2010). However, up to date, only one study has applied ABC to estimate global dN/dS (Wilson et al. 2009).

6.6 Practical Examples

We performed global and local dN/dS estimation in 3 datasets from a highly recombining virus, HIV-1. The dN/dS values were estimated with the *Datamonkey* web server. Estimations were performed by ignoring and considering recombination. In the first case we reconstructed a neighbour joining (NJ) (Saitou and Nei 1987) tree for the whole alignment and estimated the global dN/dS and the number of PSS and negatively selected sites (NSS) using the SLAC method (Kosakovsky Pond and Frost 2005a). Under the second strategy, we first detected recombination breakpoints using GARD (*Datamonkey* web server) and reconstructed a NJ tree for each recombinant fragment. Then, we computed dN/dS, number of PSS and number of NSS using the SLAC method for each fragment. Table 6.1 shows the results under both strategies. Overall, the global dN/dS estimates do not vary much when recombination is ignored. However, notice how the number of estimated PSS and NSS often increases when recombination is ignored. Similar effects have also been observed by Castillo-Ramirez et al. (2011).

6.7 Concluding Remarks

Recombination affects the estimation of dN/dS values at particular sites. Because most methods used to estimate

Table 6.1. Global and local *dN/dS* estimation by considering/ignoring recombination from real data. The 3 datasets belong to HIV-1 and were downloaded from the *PopSet* database. Alignment length is shown in number of nucleotides. Recombination breakpoints were detected with the *GARD* method, which is implemented in the *Datamonkey* web server. Molecular adaptation was estimated using the *SLAC* method, which is also implemented in the *Datamonkey* web server. PSS and NSS indicate positively and negatively selected sites at the default 90% of significance.

Entry	Sample size, alignment length	Number of detected recombination events	Ignoring recombination			Considering recombination			Reference
			Global *dN/dS*	Number of PSS	Number of NSS	Global *dN/dS*	Number of PSS	Number of NSS	
1	37, 1002	3	0.310	14	100	0.300	9	96	(de Souza et al. 2008)
2	35, 423	1	0.818	9	12	0.775	7	12	(Ryan et al. 2007)
3	23, 432	1	0.570	7	23	0.554	6	24	(Zhong et al. 2003)

121

dN/dS are based on phylogenetic reconstructions, recombination therefore plays an important role on the inference of natural selection. Obviously, this is especially relevant for those studies trying to detect positive selection across recombining genomes (e.g., Nielsen and Yang 1998; Zanotto et al. 1999; de Oliveira et al. 2004; Bustamante et al. 2005; Clark et al. 2007; Nielsen et al. 2007; Castillo-Ramirez et al. 2011).

Acknowledgements

This study was partially supported by the Spanish Government through the "Juan de la Cierva" fellowship, JCI-2011-10452 to M.A., and grants BIO2007-61411 and BFU2009-08611 to D.P.

References

Anisimova, M., R. Nielsen and Z. Yang. 2003. Effect of recombination on the accuracy of the likelihood method for detecting positive selection at amino acid sites. Genetics 164: 1229–36.

Arenas, M. and D. Posada. 2010. Coalescent simulation of intracodon recombination. Genetics 184: 429–37.

Beaumont, M.A. 2010. Approximate Bayesian Computation in Evolution and Ecology. Annu Rev Ecol Evol Syst 41: 379–405.

Beaumont, M.A., W. Zhang and D.J. Balding. 2002. Approximate Bayesian computation in population genetics. Genetics 162: 2025–35.

Behura, S.K. and D.W. Severson. 2013. Nucleotide substitutions in dengue virus serotypes from Asian and American countries: insights into intracodon recombination and purifying selection. BMC Microbiol 13: 37.

Bozek, K. and T. Lengauer. 2010. Positive selection of HIV host factors and the evolution of lentivirus genes. BMC Evol Biol 10: 186.

Bustamante, C.D., A. Fledel-Alon, S. Williamson, R. Nielsen, M.T. Hubisz, S. Glanowski, D.M. Tanenbaum, T.J. White, J.J. Sninsky, R.D. Hernandez, D. Civello, M.D. Adams, M. Cargill and A.G. Clark. 2005. Natural selection on protein-coding genes in the human genome. Nature 437: 1153–7.

Castillo-Ramirez, S., S.R. Harris, M.T. Holden, M. He, J. Parkhill, S.D. Bentley and E.J. Feil. 2011. The impact of recombination on dN/dS within recently emerged bacterial clones. PLoS Pathog 7: e1002129.

Chen, J. and Y. Sun. 2011. Variation in the analysis of positively selected sites using nonsynonymous/synonymous rate ratios: an example using influenza virus. PLoS One 6: e19996.

Clark, A.G., M.B. Eisen, D.R. Smith, C.M. Bergman, B. Oliver, T.A. Markow, T.C. Kaufman, M. Kellis, W. Gelbart, V.N. Iyer and D.A. Pollard. 2007. Evolution of genes and genomes on the Drosophila phylogeny. Nature 450: 203–18.

De Oliveira, T., M. Salemi, M. Gordon, A.M. Vandamme, E.J. Van Rensburg, S. Engelbrecht, H.M. Coovadia and S. Cassol. 2004. Mapping sites of positive selection and amino acid diversification in the HIV genome: an alternative approach to vaccine design? Genetics 167: 1047–58.

De Souza, A.C., C.M. De Oliveira, C.L. Rodrigues, S.A. Silva and J.E. Levi. 2008. Short communication: Molecular characterization of HIV type 1 BF pol recombinants from Sao Paulo, Brazil. AIDS Res Hum Retroviruses 24: 1521–5.

Kim, P.M., J.O. Korbel and M.B. Gerstein. 2007. Positive selection at the protein network periphery: evaluation in terms of structural constraints and cellular context. Proc Natl Acad Sci U S A 104: 20274–9.

Kosakovsky Pond, S.L. and S.D. Frost. 2005a. Datamonkey: rapid detection of selective pressure on individual sites of codon alignments. Bioinformatics 21: 2531–2533.

Kosakovsky Pond, S.L. and S.D. Frost. 2005b. Not so different after all: a comparison of methods for detecting amino Acid sites under selection. Mol Biol Evol 22: 1208–22.

Kosakovsky Pond, S.L., S.D. Frost and S.V. Muse. 2005. HYPHY: Hypothesis testing using phylogenies. Bioinformatics 21: 676–679.

Kosakovsky Pond, S.L., D. Posada, M.B. Gravenor, C.H. Woelk and S.D. Frost. 2006a. Automated phylogenetic detection of recombination using a genetic algorithm. Molecular Biology and Evolution 23: 1891–1901.

Kosakovsky Pond, S.L., D. Posada, M.B. Gravenor, C.H. Woelk and S.D. Frost. 2006b. GARD: a genetic algorithm for recombination detection. Bioinformatics 22: 3096–8.

Kryazhimskiy, S., G.A. Bazykin and J. Dushoff. 2008. Natural selection for nucleotide usage at synonymous and nonsynonymous sites in influenza A virus genes. J Virol 82: 4938–45.

Martin, D.P., P. Lemey, M. Lott, V. Moulton, D. Posada and P. Lefeuvre. 2010. RDP3: a flexible and fast computer program for analyzing recombination. Bioinformatics 26: 2462–3.

Martin, D.P., P. Lemey and D. Posada. 2011. Analysing recombination in nucleotide sequences. Mol Ecol Resour 11: 943–55.

Nei, M. and T. Gojobori. 1986. Simple method for estimating the numbers of synonymous and nonsynonymous nucleotide substitutions. Mol Biol Evol 3: 418–426.

Nielsen, R., I. Hellmann, M. Hubisz, C. Bustamante and A.G. Clark. 2007. Recent and ongoing selection in the human genome. Nat Rev Genet 8: 857–68.

Nielsen, R. and Z. Yang. 1998. Likelihood models for detecting positively selected amino acid sites and applications to the HIV-1 envelope gene. Genetics 148: 929–36.

Perez-Losada, M., D.V. Jobes, F. Sinangil, K.A. Crandall, M. Arenas, D. Posada and P.W. Berman. 2011. Phylodynamics of HIV-1 from a phase III AIDS vaccine trial in Bangkok, Thailand. PLoS One 6: e16902.

Perez-Losada, M., D. Posada, M. Arenas, D.V. Jobes, F. Sinangil, P.W. Berman and K.A. Crandall. 2009. Ethnic differences in the adaptation rate of HIV gp120 from a vaccine trial. Retrovirology 6: 67.

Posada, D. and K.A. Crandall. 2002. The effect of recombination on the accuracy of phylogeny estimation. J Mol Evol 54: 396–402.

Ryan, C.E., J. Gare, S.M. Crowe, K. Wilson, J.C. Reeder and R.B. Oelrichs. 2007. The heterosexual HIV type 1 epidemic in Papua New Guinea is dominated by subtype C. AIDS Res Hum Retroviruses 23: 941–4.

Saitou, N. and M. Nei. 1987. The neighbor-joining method: a new method for reconstructing phylogenetic trees. Molecular Biology and Evolution 4: 406–425.

Schierup, M.H. and J. Hein. 2000a. Consequences of recombination on traditional phylogenetic analysis. Genetics 156: 879–891.

Schierup, M.H. and J. Hein. 2000b. Recombination and the molecular clock. Molecular Biology and Evolution 17: 1578–1579.

Shriner, D., D.C. Nickle, M.A. Jensen and J.I. Mullins. 2003. Potential impact of recombination on sitewise approaches for detecting positive natural selection. Genetical Research 81: 115–121.

Wilson, D.J., E. Gabriel, A.J. Leatherbarrow, J. Cheesbrough, S. Gee, E. Bolton, A. Fox, C.A. Hart, P.J. Diggle and P. Fearnhead. 2009. Rapid evolution and the importance of recombination to the gastroenteric pathogen Campylobacter jejuni. Mol Biol Evol 26: 385–97.

Wiuf, C. and D. Posada. 2003. A coalescent model of recombination hotspots. Genetics 164: 407–17.

Yang, Z. 1999. PAML: a phylogenetic analysis by maximum likelihood. Version 2.0e. University College London, London.

Yang, Z. and R. Nielsen. 2002. Codon-substitution models for detecting molecular adaptation at individual sites along specific lineages. Mol Biol Evol 19: 908–17.

Yang, Z., R. Nielsen, N. Goldman and A.-M.K. Pedersen. 2000. Codon-substitution models for heterogeneous selection pressure at amino acid sites. Genetics 155: 431–449.

Zanotto, P.M., E.G. Kallas, R.F. De Souza and E.C. Holmes. 1999. Genealogical evidence for positive selection in the nef gene of HIV-1. Genetics 153: 1077–89.

Zhong, P., B.U. S, F. Konings, M. Urbanski, L. Ma, L. Zekeng, L. Ewane, L. Agyingi, M. Agwara, Saa, Z.E. Afane, T. Kinge, S. Zolla-Pazner and P. Nyambi. 2003. Genetic and biological properties of HIV type 1 isolates prevalent in villagers of the Cameroon equatorial rain forests and grass fields: further evidence of broad HIV type 1 genetic diversity. AIDS Res Hum Retroviruses 19: 1167–78.

Chapter 7

Why Proteins Evolve at Different Rates: The Determinants of Proteins' Rates of Evolution

David Alvarez-Ponce

7.1 Introduction

Proteins undergo changes in their amino acid sequences over evolutionary time, as a result of the accumulation of nonsynonymous mutations in their encoding genes. Genes and proteins act as "molecular clocks", accumulating changes at a relatively constant rate, as mutations occur with a certain probability each time a nucleotide is replicated. From the very beginning of molecular evolution studies, it became apparent that different proteins evolve at very different rates, each evolving according to its own "molecular clock". For instance, in their early studies Zuckerkandl and Pauling (1965) already noted that the high rate of evolution of hemoglobin was "spectacularly

Assistant Professor, Department of Biology, University of Nevada, Reno, USA.
Email: dap@unr.edu

at variance" with the high degree of conservation of cytochrome *c*. The subsequent accumulation of molecular data for other proteins revealed a huge diversity in proteins' rates of evolution. For instance, using DNA sequence data from 36 genes in different mammalian species, Li et al. (1985) observed that the gene encoding interferon γ had accumulated nonsynonymous mutations at a rate that was ~700-fold higher than that for the gene encoding histone H4. The current availability of the complete genomes of multiple organisms now allows us to study proteins' rates of evolution at an unprecedented scale. If we take any two genomes (e.g., those of human and mouse), and compare the proteins encoded by each pair of orthologous genes, some proteins will be very similar (or even identical) in both organisms, whereas others will exhibit several amino acid differences (see Figure 7.1), in spite of the fact that all pairs of orthologs diverged over the same amount of time (i.e., the time elapsed since the divergence of the two compared species).

Why do certain proteins accumulate mutations at a high velocity, whereas others remain virtually immutable over long evolutionary periods? This question has attracted the interest of evolutionary biologists for decades and is still a highly active area of research. Zuckerkandl and Pauling (1965) attributed the differences in the rates of evolution of hemoglobin and cytochrome *c* to their being subject to different levels of selective constraint. Kimura and Ohta (1974) deduced from the neutral theory of molecular evolution (Kimura 1968, 1983) that less functionally important proteins (or protein domains) should evolve at faster rates than more important ones. In addition, it was postulated that levels of selective constraint acting on a protein depended on the proportion of amino acids involved in its function (i.e., its "functional density") (Zuckerkandl 1976). Out of the entire set of amino acid changes that a

127

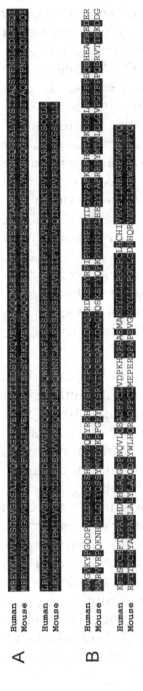

Figure 7.1. Sequence alignments of human proteins RAP1B (A) and TEX37 (B) and their mouse orthologs. RAP1B is identical in both organisms, whereas TEX37 has accumulated several amino acid replacements since both organisms diverged ~75 million years ago. Identical amino acids in both species are represented in a black background.

protein can undergo, many will affect its function, thus decreasing the fitness of the organisms carrying them; such mutations will likely be removed by purifying (or negative) selection. A fraction of the possible changes, however, will have little or no effect on the protein's function (and hence on the organism's fitness), and thus will possibly be fixed in the population. Proteins with a low functional density are expected to be less selectively constrained, and thus to evolve at higher rates. However, despite the plausibility of the notion that proteins' rates of evolution are mainly determined by their importance and/or functional density, the hypothesis is difficult to test, as these parameters are very difficult to measure experimentally.

In the last years, the emergence of genome-scale datasets containing measures of a number of characteristics for an important fraction of genes from model organisms has allowed researchers to identify a long list of factors that correlate with rates of evolution. Perhaps surprisingly, the prevailing view is that factors that relate to proteins' importance and/or functional density, such as functional category (Pál et al. 2001; Rocha and Danchin 2004; Greenberg et al. 2008; Alvarez-Ponce and Fares 2012), number of functions (Wilson et al. 1977; Salathé et al. 2006; Podder et al. 2009), essentiality for the organism's survival (Hurst and Smith 1999; Jordan et al. 2002; Rocha and Danchin 2004), dispensability (i.e., fitness effect upon gene knockout; Hirsh and Fraser 2001; Yang et al. 2003; Wall et al. 2005; Zhang and He 2005), and number of protein–protein interactions (Fraser et al. 2002), appear to be only relatively weak predictors of rates of evolution. On the other hand, factors that have little to do with function, such as patterns and levels of gene expression (Duret and Mouchiroud 2000; Pál et al. 2001b) (Figure 7.2), seem to be the strongest determinants of levels of selective constraint—indeed, some reports suggest that,

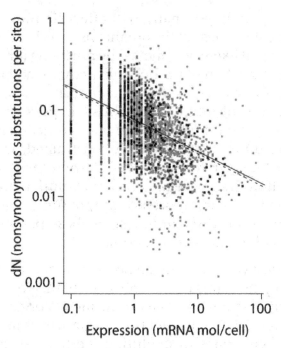

Figure 7.2. Correlation between gene expression level and degree of nonsynonymous divergence in yeast. Figure from Drummond et al. (2005), reproduced with permission from the authors and the publisher. Copyright 2005 National Academy of Sciences, U.S.A.

in yeasts, gene expression levels may account for more than 30% of the variability of rates of evolution (Drummond et al. 2005). Not all analyses, however, concur in this conclusion, and some suggest that certain factors may be as important as expression in determining proteins' rates of evolution (e.g., Liao et al. 2006; Ingvarsson 2007; Plotkin and Fraser 2007; Alvarez-Ponce 2012).

In addition to the variable strength of purifying selection acting on amino acid sequences, at least three other factors can contribute to the variability of proteins' rates of evolution. First, due to the stochastic nature of mutations, the number of differences between two homologous proteins will be subject to random variation, particularly if they are very closely related. Second, positive selection can increase proteins' rates of evolution above neutrality levels. Third, DNA mutation rates are variable across the genome, with certain chromosome regions exhibiting particularly high mutation rates. As a result, genes located at regions with high mutation rates (and, as a result, proteins encoded by these genes) are expected to evolve at a fast rate (Wyckoff et al. 2005). How to tell apart the effect of natural selection from that of genomic mutation rates? One method generally used to discount the effect of DNA mutation rates on the rate of evolution of a given protein is dividing the degree of nonsynonymous divergence (d_N, the estimated number of nonsynonymous substitutions per nonsynonymous position) by the degree of synonymous divergence (d_S, the number of synonymous substitutions per synonymous position). Synonymous positions are generally assumed to evolve in an approximately neutral fashion, and hence their rate of evolution is assumed to reflect mutation rates, and can be used as a benchmark against which rates of nonsynonymous evolution can be compared. For genes evolving neutrally (e.g., a pseudogene, wherein mutations have no effect on fitness, and hence are not filtered by natural selection), nonsynonymous and synonymous positions are expected to evolve at similar rates, resulting in values of d_N/d_S close to 1. In highly constrained genes, a high fraction of nonsynonymous mutations will be removed by natural selection, resulting in values of d_N/d_S much lower than 1. The nonsynonymous to synonymous divergence ratio (d_N/d_S) can thus be taken as a measure of levels of selective constraint (or the strength of purifying selection) acting on a given protein, with lower d_N/d_S ratios being indicative of higher levels of selective constraint. Throughout this chapter, as is common practice in scientific literature, the term "evolutionary rate" refers indistinctly to proteins' rates of evolution (i.e., the number of amino acid replacements per amino acid position and time unit) and to the d_N/d_S ratio—indeed, high d_N/d_S ratios often result in high rates of protein evolution.

In this chapter, the current knowledge on the determinants of proteins' rates of evolution is reviewed. For each factor, the observed trends are described, along with the mechanisms that have been proposed to explain them. The chapter ends with a discussion of the difficulties of establishing a factor as a *bona fide* determinant of rates of evolution, along with the multivariate analysis techniques commonly used to rule out the effects of potentially confounding factors.

7.2 Gene Expression Level and Breadth Seem to be the Main Determinants of Levels of Selective Constraint

It has been repeatedly shown that genes expressed at high levels (as inferred from the presence of a high number of mRNA molecules in the cell) exhibit lower d_N/d_S ratios than those expressed at low levels (see Figure 7.2). This trend has been observed all across life, from bacteria (Rocha and Danchin 2004; Drummond et al. 2006; Dötsch et al. 2010), to viruses (Pagán et al. 2012), yeasts (Pál et al. 2001b; Drummond et al. 2005; Drummond et al. 2006), paramecia (Gout et al. 2010), algae (Popescu et al. 2006; Chang and Liao 2013), plants (Renaut et al. 2012; Wright et al. 2004; Ingvarsson 2007; Slotte et al. 2011; Yang and Gaut 2011; Alvarez-Ponce and Fares 2012; Paape et al. 2013), nematodes (Krylov et al. 2003), insects (Lemos et al. 2005), and mammals (Subramanian and Kumar 2004; Alvarez-

Ponce 2012). Indeed, in unicellular organisms, expression level seems to be the main determinant of levels of selective constraint.

A number of models have been proposed to explain the observed negative relationship between genes' levels of expression and proteins' rates of evolution. The translational efficiency hypothesis (Akashi 2001; Akashi 2003) is based on the observation that highly expressed genes tend to display a highly biased codon usage. Among the set of synonymous codons that code for any given amino acid, some (usually those that are translated at a higher speed and with greater accuracy) are preferred over the others (a phenomenon known as codon usage bias; for review, see Hershberg and Petrov 2008). Highly expressed genes exhibit a particularly strong codon bias, using preferred codons very often, which facilitates their fast and accurate translation, whereas lowly expressed genes can make use of a wider spectrum of codons. Mutations at the coding sequences (CDSs) of highly expressed genes, whether synonymous or nonsynonymous, will likely provoke changes from preferred codons to unpreferred (i.e., suboptimal) ones, thus reducing the speed at which the gene is translated, and/or increasing translation error rates. This is expected to result in highly expressed genes (and hence the encoded proteins) being highly constrained. The problems of this model include the facts that selection on codon usage is too weak to explain the expression level-evolutionary rate anticorrelation, and that this correlation seems to be universal, being observed even in organisms with little codon bias (Drummond et al. 2005; Rocha 2006).

The functional loss hypothesis (Rocha and Danchin 2004) relies on the assumption that each protein molecule has roughly the same contribution to the organism's fitness. Genes expressed at high levels would hence be more important for the organism's fitness. Therefore, gene mutations decreasing the functional efficiency of the encoded proteins would have higher deleterious effects if they affect highly expressed genes, as a higher number of protein molecules would be affected. One of the problems of this hypothesis is that certain proteins (such as DNA polymerases and most regulatory proteins) are present in low numbers in the cell (as their target molecules are also present at low concentrations), yet their function is crucial for the organism (Drummond et al. 2005; Rocha 2006).

The translational robustness hypothesis (Drummond et al. 2005) proposes that highly expressed proteins are under stronger selective pressure to tolerate translation errors, thereby making them more selectively constrained at the sequence level. The cell's translational machinery has a relatively high error rate, which means that ~20% of all yeast translated protein molecules carry one or more erroneous amino acids. As a result, some defective proteins will not fold or function properly, and in some cases will aggregate, thus becoming toxic to the cell. This phenomenon would result in a significant amount of energy being wasted in the synthesis of proteins that are non-functional, or even toxic (Geiler-Samerotte et al. 2010). However, according to the translational robustness hypothesis, certain proteins may have evolved the ability to appropriately fold and function despite carrying some mistranslation errors (i.e., they may be robust to mistranslation). Such proteins may thus be highly constrained at the sequence level, in order to preserve their robustness to mistranslation. This effect is expected to be stronger in highly translated genes (indeed, misfolding and aggregation of lowly translated proteins may be relatively unproblematic), thus resulting in a negative correlation between translation rate (approximated through mRNA levels) and proteins' rates of evolution.

The translational robustness hypothesis has received support from both genomic observations (Drummond et al. 2005; Drummond et al. 2006; Drummond and Wilke 2008) and theoretical population genetics models (Wilke and Drummond 2006), being the prevailing model nowadays, to explain the expression level-evolutionary rate anticorrelation. According to this hypothesis, the main determinant of a protein's evolutionary rate would be its translation rate, rather than its abundance (which depends on the balance between translation and degradation). Consistent with this prediction, evolutionary rates seem to correlate better with mRNA abundances (which were used as a proxy for translation rate, as unfortunately no genome-scale translation rate dataset is currently available) than with protein abundances (Drummond et al. 2005). Furthermore, a principal components regression analysis taking into account several correlates of rates of evolution in yeast showed that the first component (mostly dominated by codon bias, mRNA abundance and protein abundance, which may well be a reflection of translation rate) accounts for 43% of the variance in proteins' rates of evolution, whereas all the other components account for less than 1% (Drummond et al. 2006)—this result, however, might be biased by the different degrees of accuracy with which the different factors were measured (Plotkin and Fraser 2007). Further support for the translational robustness hypothesis is provided by the observation that the correlation between expression level and evolutionary rate is particularly strong in tissues composed of neurons, whose delicate structure and long lifetime make them particularly sensitive to protein misfolding (Drummond and Wilke 2008; Tuller et al. 2008).

It should be noted, however, that despite the likely explanatory power of the translational robustness hypothesis, it probably does not completely explain the strength of the correlation between genes' expression levels and proteins' rates of evolution. Other recently proposed effects, such as highly expressed genes being under stronger selective pressure to have a high mRNA folding strength (Park et al. 2013), and/or a low rate of unspecific protein–protein interactions (Levy et al. 2012; Yang et al. 2012), may also contribute to the correlation. Another recently proposed model argues that mutations affecting proteins' function will have a stronger fitness effect for highly expressed genes, owing to the energetic costs of synthesising non-functional, or partially functional, proteins (Cherry 2010; Gout et al. 2010).

In multicellular organisms, expression breadth (i.e., the number of different tissues in which a gene is expressed), also seems to be a major determinant, of proteins' rates of evolution, with widely expressed genes being more selectively constrained than genes expressed in a narrow set of tissues (Duret and Mouchiroud 2000; Wright et al. 2004; Zhang and Li 2004; Liao et al. 2006; Pál et al. 2006; Ingvarsson 2007; Alvarez-Ponce and Fares 2012). Indeed, some analyses have suggested that, once the effect of expression breadth is discounted, the effect of expression level is only residual, or even non-existing (Pál et al. 2006; Ingvarsson 2007). Other analyses, on the contrary, suggest a greater influence of expression levels (Subramanian and Kumar 2004). The tendency of broadly expressed genes to be more selectively constrained could be the result of their being more pleiotropic. For instance, genes expressed in a wide range of tissues may be involved in a greater array of biochemical processes or pathways, may have to function in a wider range of biochemical environments, and/or may have to interact with more proteins (Kuma et al. 1995; Duret and Mouchiroud 1999).

7.3 The Surprisingly Weak Effect of Gene Essentiality and Dispensability on Proteins' Rates of Evolution

Loss of certain genes (e.g., by knockout) results in the organism's death, whereas loss of other genes is compatible with the organism's survival. Early models, based on the neutral theory of molecular evolution, predicted that essential genes should exhibit lower rates of evolution than non-essential ones (Wilson et al. 1977). As soon as large-scale genomic and functional datasets made it possible, evolutionary biologists sought out to test this hypothesis. Multiple analyses have shown that, indeed, essential proteins evolve slower than non-essential ones; however, the difference between both groups seems to be rather small. Indeed, in some early analyses the differences even vanished once potential confounding factors (such as expression level, duplicability, or positive selection) were controlled for (Hurst and Smith 1999; Rocha and Danchin 2004). Several subsequent analyses, however, indicate that essentiality has an effect on proteins' rates of evolution that is independent of confounding factors (Jordan et al. 2002; Castillo-Davis and Hartl 2003; Zhang and He 2005; Liao et al. 2006; Wolf et al. 2006; Larracuente et al. 2008; Dötsch et al. 2010). In addition to direct evidence, the notion that essential genes are more selectively constrained is also supported by the observation that genes that are reticent to be lost during evolution (probably as a result of their essentiality) exhibit lower rates of evolution than genes that are prone to be lost (Krylov et al. 2003).

Another related parameter that exhibits a weak correlation with proteins' rates of evolution is gene dispensability, i.e., the inverse of the fitness effect of the loss of a given gene (Hirsh and Fraser 2001). A gene's dispensability can be estimated, for example, from the decrease in growth rate of a unicellular organism that results from loss of the gene. The association of dispensability with rates of evolution has been somewhat controversial, with some reports suggesting that dispensability has no effect on proteins' rates of evolution once confounding factors are factored out (Pál et al. 2003; Yang et al. 2003; Drummond et al. 2006), and others pointing out to a weak, albeit measurable, effect that is independent of known confounding factors (Chen and Xu 2005; Wall et al. 2005; Zhang and He 2005; Kim and Yi 2007; Plotkin and Fraser 2007; Wang and Zhang 2009).

A number of hypotheses have been proposed to explain the weakness of the relationship between evolutionary rates and essentiality and dispensability. First, it has been argued that essentiality and dispensability measures obtained under laboratory conditions may be relatively poor predictors of the importance of a gene in nature (Pál et al. 2006). Indeed, some genes can be unessential under the favourable conditions of a laboratory, but may be essential under certain natural, suboptimal, conditions. However, analyses by Wang and Zhang (2009), using sets of dispensabilities measured or computationally predicted under a wide diversity of conditions, suggest that this may not be the reason why a gene's importance and rate of evolution are so weakly correlated: under none of the tested conditions dispensabilities were a strong predictor of rates of evolution. Second, essentiality and dispensability measure the effect of a gene's loss; such effect, however, may be different from that of punctual mutations, which is the kind of effect estimated by d_N/d_S (Pál et al. 2006). Third, essential genes in one species may not be essential in other species. Consistently, Zhang and He (2005) observed that, on average, nonessential yeast proteins evolve 40% faster than essential ones when evolutionary rates are estimated from comparison of closely related yeast species, but that the difference drops down to 10% when more distant yeast species are used, suggesting a relatively low degree of conservation of gene essentialities across taxa.

7.4 The Structural Determinants of Proteins' Rates of Evolution

Proteins adopt a huge diversity of three-dimensional structures, and a number of aspects of these structures have been shown to affect their rates of evolution. A protein's "designability" is defined as the number of possible sequences that are compatible with the native structure (Li et al. 1996; Kussell 2005). Highly designable proteins can tolerate a high number of mutations, and hence are expected to evolve at fast rates. Indeed, different measures of designability, such as contact density (the average number of contacts per residue) and protein stability, positively correlate with proteins' rates of evolution (Bloom et al. 2006a; Bloom et al. 2006b; Zhou et al. 2008; Toll-Riera et al. 2012).

Amino acids within a protein contribute differentially to its structure and function, and hence are expected to be subject to different selective pressures. For instance, amino acids that locate at the protein's surface (i.e., those with a high solvent accessibility) evolve twice as fast as those that are buried within the protein, which may be under stronger selective constraint to maintain the structure of the protein (Thorne et al. 1996; Goldman et al. 1998; Bustamante et al. 2000; Bloom et al. 2006a; Lin et al. 2007; Liu et al. 2008; Conant 2009; Franzosa and Xia 2009; Toll-Riera et al. 2012). Among amino acids locating at the surface, those involved in inter-molecular interactions are relatively highly constrained (Ingram 1961; Dickerson 1971; Wilson et al. 1977; Kisters-Woike et al. 2000; Mintseris and Weng 2005). Furthermore, amino acids' rates of evolution negatively correlate with the number of amino acids of the same protein with which they interact (Toft and Fares 2010). Finally, intrinsically disordered regions (i.e., those lacking a rigid three-dimensional structure under physiological conditions) tend to evolve much faster than ordered regions (Brown et al. 2002; Liu et al. 2008), and among ordered regions, those adopting different secondary structures (α-helices, β-sheets and loops) evolve at different rates, with β-sheets being slightly more conserved (Thorne et al. 1996; Goldman et al. 1998; Bloom et al. 2006a).

7.5 The Complex Relationship Between Proteins' Lengths and Rates of Evolution

A number of studies have reported correlations between proteins' lengths and their rates of evolution. Indeed, some analyses have found a strong influence of this factor, in some cases even surpassing that of expression level and breadth (Liao et al. 2006; Ingvarsson 2007). Other analyses, on the contrary, have suggested little or no effect of proteins' length on rates of evolution (Drummond et al. 2006; Alvarez-Ponce 2012). More strikingly, different analyses have suggested different signs of the correlation between protein length and rates of evolution: some reports indicate that short proteins tend to be more selectively constrained (Marais and Duret 2001; Lemos et al. 2005; Bloom et al. 2006a; Drummond et al. 2006; Ingvarsson 2007; Alvarez-Ponce and Fares 2012; Chang and Liao 2013), whereas others point to the opposite trend (Liao et al. 2006; Larracuente et al. 2008). Furthermore, the correlation between protein length and evolutionary rate is stronger among short proteins (Bloom

140

et al. 2006a). These contrasting observations indicate that no general rule governs the relationship between proteins' lengths and rates of evolution. The mechanisms underlying the complex dependence between protein length and evolutionary rates remain largely unclear, although both the Hill-Robertson effect and the non-monotonic association between expression level and protein length may play relevant roles.

Linkage between surrounding genomic positions can reduce the efficacy of natural selection. Fixation of a beneficial variant at a given position may entail fixation of deleterious mutations at linked positions, and conversely, elimination of a deleterious mutation may entail elimination of a linked advantageous one. Likewise, adaptive mutations at linked positions can compete for fixation (simultaneous fixation of mutations at two linked positions is not possible, unless both mutations belong to the same haplotype). This interference, known as the Hill-Robertson effect (Hill and Robertson 1966), is expected to be stronger for longer genes, in which many sites can be subject to selection simultaneously (Ingvarsson 2007), and can cause genes to experience a decreased efficacy of natural selection (either positive or negative). This can result in a correlation (positive or negative, depending on whether negative or positive selection is the predominant selective pressure). On the other hand, genes encoding longer proteins tend to contain a high number of introns (e.g., Larracuente et al. 2008), and introns, by allowing recombination, reduce Hill-Robertson interference (Comeron and Kreitman 2000, 2002).

The relationship between proteins' length and rates of evolution may be further complicated by the relationship between these variables and expression level. As developed above, rates of evolution are strongly negatively correlated with gene expression levels (Pál et al. 2001b). Hence, an association between protein length and gene expression level might explain the association between proteins' lengths and rates of evolution. If protein length and expression level are positively correlated, then protein length and evolutionary rate would be expected to exhibit a negative correlation; conversely, a negative protein length-expression level correlation would result in a positive protein length-evolutionary rate correlation. Remarkably, the relationship between protein length and levels of gene expression is nonmonotonic, being positive for lowly expressed genes and negative for highly expressed genes (Carmel and Koonin 2009). Although some of previous analyses linking protein length to rates of evolution have controlled for expression levels (Ingvarsson 2007; Larracuente et al. 2008), the nonmonotonic nature of the relationship between expression level and protein length has not been taken into account.

In summary, the relationship between proteins' lengths and rates of evolution is complex, which is perhaps not surprising upon consideration of the complex interplay of factors that may shape it. Further work is warranted to understand this relationship.

7.6 The Time-dependent Effect of Genes' Duplication on Proteins' Rates of Evolution

Gene duplications are relatively frequent events that have a great impact on genome evolution, and are thought to be at the basis of genome complexity and the evolution of new functions (Ohno 1970). They are the result of unequal crossing-over during meiosis (which leads to tandem or closely located duplicates), retrotranspositions (which result in the ancestral gene that conserves its position and a processed, intronless retrogene that inserts into another genomic location), segmental duplications (which result in duplicated chromosomal regions of variable size), and whole genome duplications (which result in the simultaneous duplication of all genes in a genome).

Immediately after gene duplication, both copies are identical or, at least, encode identical proteins. Usually, due to this redundancy, one of the copies undergoes pseudogenization and is eventually lost. However, under certain conditions, both copies can be retained. Three models, proposed by Ohno (1970), are commonly invoked to explain the retention of duplicates (for review, see Hahn 2009; Innan and Kondrashov 2010). Each pair of duplicates can evolve under a different model or combination of models, and each model is expected to result in different patterns on the rates of evolution of the resulting copies. According to the neofunctionalization model, one of the copies retains the functions of the ancestral gene thanks to purifying selection, whereas the redundant copy experiences a period of neutral evolution and, as a result, can acquire new functions and potentially, owing to these new functions, undergo positive selection. This results in a rate acceleration that affects specifically the neofunctionalized copy; the copy maintaining the ancestral functions, on the contrary, will be subject to selective pressures that are equivalent to the ancestral gene. In the subfunctionalization model, the functions of the ancestral gene are divided among the two resulting duplicates, with each duplicate retaining a subset of the functions of the ancestral gene (note that "functions" can refer to either the biochemical functions of the encoded proteins, or the tissues in which a gene is expressed). This model predicts a relaxation of purifying selection in both gene copies, as each of them will perform less functions than the ancestral gene; the rates of evolution of both duplicates, nevertheless, need not be identical (see He and Zhang 2005). Finally, the gene conservation model involves the maintenance of all functions of the ancestral gene in the duplicates; in this case, both duplicates are maintained, e.g., because the resulting increase in expression levels is advantageous. Under this model, rates of evolution are expected to be equivalent for the ancestral gene and for both duplicates.

Several works have reported accelerated rates of evolution immediately after gene duplication (Lynch and Conery 2000; Van de Peer et al. 2001; Kondrashov et al. 2002; Nembaware et al. 2002; Scannell and Wolfe 2008; Panchin et al. 2010; Pegueroles et al. 2013). The initial increase in proteins rates of evolution is, however, subsequently attenuated; i.e., after an initial rate of evolution acceleration, evolutionary rates seem to be reverted to pre-duplication levels. Some analyses, furthermore, note an asymmetry in the rates of evolution of both gene copies (i.e., one copy evolving faster than the other) in an important fraction of duplicated pairs (5%–30%, a figure that might have been underestimated due to limited statistical power; Lynch and Katju 2004) (Conant and Wagner 2003; Zhang et al. 2003; Kellis et al. 2004; Chain and Evans 2006; Scannell and Wolfe 2008; Panchin et al. 2010). This asymmetry seems to be specific to distant duplicates, including retrogenes, not affecting tandem duplicates (Cusack and Wolfe 2007; Jun et al. 2009). Furthermore, the fastest-evolving gene tends to be the daughter copy, i.e., the one that moves away from the location of the parental gene, which tends to undergo positive selection and to change its expression patterns (Han et al. 2009; Pegueroles et al. 2013). The parent gene, on the contrary, tends to evolve at a rate equivalent to that of the ancestral gene, consistent with the neofunctionalization model (Pegueroles et al. 2013).

144

Despite the increase in proteins' rates of evolution immediately after gene duplication, comparison of duplicated and single-copy (singleton) genes reveals that duplicated genes are, on average, more selectively constrained than singleton genes (Nembaware et al. 2002; Yang et al. 2003; Davis and Petrov 2004; Jordan et al. 2004; Yang and Gaut 2011). This may be the result of functionally important genes being both more selectively constrained and more likely to be retained as duplicates; indeed, *Caenorhabditis elegans* and *Saccharomyces cerevisiae* duplicated genes were 25%–50% more constrained prior to duplication than genes that did not duplicate (Davis and Petrov 2004). Furthermore, proteins' rates of evolution negatively correlate with the number of paralogs of the encoding genes (Jordan et al. 2004; Alvarez-Ponce 2012).

7.7 The Buffering Effect of Chaperones Allows Proteins to Evolve Faster

Molecular chaperones are proteins that, among other functions, help other proteins to fold correctly into their functional conformation (for review, see Bogumil and Dagan 2012; Henderson et al. 2013). Several lines of evidence indicate that chaperones can mask certain deleterious mutations that, in the absence of chaperones, would manifest in the form of aberrant phenotypes. Consistently, inhibition of Hsp90 in *Drosophila* and *Arabidopsis* results in phenotypic deformities, probably as a result of misfolding of Hsp90 clients (Rutherford and Lindquist 1998; Queitsch et al. 2002). Furthermore, the fitness decrease resulting from large mutational loads in a hypermutator *Escherichia coli* strain can be largely recovered by overexpression of the GroE operon (Fares et al. 2002), and cells evolving under increased mutation rates, or under inefficient natural selection, both of which result in increased fixation of deleterious mutations, tend to over-express chaperones (Baumann et al. 1996; Maisnier-Patin et al. 2005).

These observations led to the hypothesis that interaction of proteins with chaperones, by increasing the fraction of mutations that are neutral, increases proteins' evolvability (Rutherford 2003; Tokuriki and Tawfik 2009b; for review, see Bogumil and Dagan 2012). Consistently, experimental evolution analyses involving a few clients of GroE showed that they accumulated more changes when the chaperone was over-expressed (Tokuriki and Tawfik 2009a). Furthermore, comparative genomics analyses have shown that bacterial GroE clients evolve faster than proteins that fold independently of GroE (Bogumil and Dagan 2010; Williams and Fares 2010), and that Hsp90 clients tend to evolve faster than their non-client paralogs (Lachowiec et al. 2013). Finally, proteins interacting with different chaperones in the yeast chaperone-client network tend to be expressed at different levels and to evolve at different rates (Bogumil et al. 2012).

7.8 Proteins Acting at Different Subcellular Compartments Evolve at Different Rates

A number of works have reported that proteins acting at the extracellular space (i.e., secreted proteins) tend to exhibit faster rates of evolution than proteins acting at the cell membrane or inside the cell (Winter et al. 2004; Julenius and Pedersen 2006; Dean et al. 2008; Cui et al. 2009; Liao et al. 2010). Such proteins also tend to be lowly connected in the protein–protein interaction network and to perform distinctive functions. Their encoding genes often exhibit signals of positive selection, are less likely to be essential, and display distinctive levels of expression (Julenius and

Pedersen 2006; Kim et al. 2007; Dean et al. 2008). However, the tendency of extracellular proteins to evolve fast seems to be independent of these factors (Julenius and Pedersen 2006; Liao et al. 2010). The reasons for this trend remain unclear, although some proposals include the possibility that extracellular proteins may evolve faster owing to their exposure to the variable extracellular environment, including pathogens and parasites, or to their presenting disulfide bridges that stabilize their structure (Hegyi and Bork 1997; Julenius and Pedersen 2006).

Transmembrane proteins tend to exhibit intermediate rates of evolution, evolving slower than extracellular proteins, but faster than intracellular ones (Cui et al. 2009; Hudson and Conant 2011). Interestingly, Heger et al. (2009) observed that the extracellular and intracellular parts of these proteins exhibit different evolutionary rates: the extracellular part evolves faster than the intracellular part, at a rate that is similar to that of secreted proteins.

Finally, among genes acting inside the cell, those acting at different subcellular compartments also seem to be subject to different selective pressures (Kim et al. 2007; Cui et al. 2009; Hudson and Conant 2011).

7.9 Position of Proteins in Molecular Networks: Comparative Genomics Meets Systems Biology

Genes and proteins do not act in isolation. Instead, they function as pieces of a complex machinery of interacting molecules. Therefore, considering a protein's interactions with other molecules can help understanding the selective pressures acting on it (for review, see Eanes 1999; Cork and Purugganan 2004; Eanes 2011).

147

The pattern of interactions between molecules in molecular systems can be represented as networks. Different kinds of networks can be considered, depending on the nature of the molecular interactions that they represent. The entire set of physical, direct interactions among an organism's proteins constitutes its interactome. An interactome can be represented as a protein–protein interaction network, whose nodes represent proteins and edges (links) represent protein–protein interactions. Likewise, the set of biochemical reactions that can occur in a cell defines its metabolome. A metabolome can be represented in the form of a metabolic network, whose nodes represent enzymes and links represent shared metabolites (i.e., the reaction product of one enzyme is a substrate for the other; Wagner and Fell 2001). Finally, in a gene regulatory network, nodes represent transcription factors, and edges represent regulatory (activating or inhibitory) relationships.

In the last years, technological advances have allowed scientists to discover molecular interactions at a genomic scale (for review, see Fares et al. 2011). As a result, we are starting to have initial drafts of the interactomes and metabolomes of a handful of model organisms (essentially, *E. coli, S. cerevisiae, Drosophila melanogaster, C. elegans, Homo sapiens*, and *Arabidopsis thaliana*) (Stark et al. 2006; Chatr-Aryamontri et al. 2013). Although currently available datasets have a relatively low quality, being subject to very high false positive and false negative rates (von Mering et al. 2002; Bader et al. 2004; Deeds et al. 2006; Kelly and Stumpf 2012), they allow us to perform a first evaluation of how proteins' rates of evolution depend on their position in molecular networks. The dependency manifests in a variety of ways.

In early studies, a number of authors hypothesised that a protein's rate of evolution should decrease as the number of molecular interactions in which it is involved increases, as interactions impose functional constraints on the involved residues (Ingram 1961; Dickerson 1971; Wilson et al. 1977). Consistent with this prediction, proteins that form part of protein complexes tend to evolve slow (Teichmann 2002), and proteins' rates of evolution negatively correlate with their numbers of described interactions (i.e., the more interactors a protein has, the more selectively constrained it is), a pattern that has been described in all available interactomes (Fraser et al. 2002; Jordan et al. 2003; Agrafioti et al. 2005; Hahn and Kern 2005; Lemos et al. 2005; Plotkin and Fraser 2007; Davids and Zhang 2008; Wang et al. 2010; Alvarez-Ponce 2012; Alvarez-Ponce and Fares 2012), as well as in a number of smaller datasets corresponding to individual well-characterized pathways (e.g., Casals et al. 2011; Fitzpatrick and O'Halloran 2012; Lavagnino et al. 2012; Invergo et al. 2013). The correlation, however, is often rather weak (but see, for instance, Plotkin and Fraser 2007; Wang et al. 2010; Alvarez-Ponce 2012), and some analyses have even suggested that it might be a by-product of highly connected genes being highly expressed (Bloom and Adami 2003; Bloom and Adami 2004), whereas others have shown that both expression levels and the number of protein–protein interactions independently affect rates of evolution (Fraser et al. 2003; Fraser and Hirsh 2004; Fraser 2005; Lemos et al. 2005; Plotkin and Fraser 2007; Alvarez-Ponce 2012; Alvarez-Ponce and Fares 2012).

The observation that the correlation between proteins' connectivities and rates of evolution is relatively weak is perhaps not surprising, considering that residues involved in interactions (directly or indirectly) may represent a small fraction of the proteins' length. Another factor that might

149

contribute to the weakness of the correlation is the low quality and incompleteness of interactomic datasets (Bader et al. 2004; Deeds et al. 2006; Kelly and Stumpf 2012), which may result in low-quality connectivity estimates. Nonetheless, in spite of the fact that, in general, connectivity is a poor predictor of rates of evolution, the correlation between connectivity and rates of evolution seems to be particularly strong for certain functional categories (Alvarez-Ponce and Fares 2012); for instance, among human proteins involved in signal transduction, connectivity correlates with d_N/d_S as well as expression breadth, and better than expression level (Alvarez-Ponce 2012).

In addition to connectivity (the number of proteins with which a protein interacts), which is a local measure of network centrality, other, more global measures of network centrality have been shown to correlate with proteins' rates of evolution, with genes acting at the centre of the network being more selectively constrained. Closeness (the inverse of the average shortest distance to all the other proteins in the protein–protein interaction network), and in particular betweenness (the number of shortest paths between all pairs of proteins of which a given protein is part; Freeman 1977), have been shown to correlate with proteins' rates of evolution better than connectivity (Hahn and Kern 2005; Alvarez-Ponce and Fares 2012). Proteins with a high closeness and/or betweenness are particularly important for the flow of information across the network; for instance, high-betweenness proteins tend to connect modules of the network that, otherwise, would rest disconnected (Ravasz et al. 2002). Closeness and betweenness are thus approximate measures of how important a protein is to control the flux of information across the network. The observation that these centrality metrics are better correlates of evolutionary rates than connectivity suggests that the effect of the network on the evolutionary rates of its components is not mediated by direct interactions alone; the relative position of proteins in the network, in a broader sense, has an effect on their evolutionary rates as well.

Further evidence of a link between proteins' position in the protein–protein interaction network and patterns of molecular evolution is provided by the observation that genes encoding interacting proteins tend to exhibit correlated evolutionary histories (for review, see Lovell and Robertson 2010). For instance, they tend to undergo co-duplication at the same evolutionary times (Fryxell 1996; Doherty et al. 2012), and to evolve at relatively similar rates (Fraser et al. 2002; Agrafioti et al. 2005; Lemos et al. 2005; Cui et al. 2009; Clark and Aquadro 2010; Alvarez-Ponce and Fares 2012). Interacting proteins may evolve at similar rates as a result of their coevolutionary dynamics; indeed, mutations in a protein may be compensated by compensatory mutations in interacting proteins. Nevertheless, other factors such as interacting proteins being expressed at similar levels, or having similar functions, might, at least partially, account for their similar rates of evolution (Wang and Lercher 2011; Clark et al. 2012).

Analysis of metabolic networks reveals similar patterns to those observed in protein–protein interaction networks (for review, see Wagner 2012). Genes encoding the most connected enzymes (e.g., those that share metabolites with a higher number of other enzymes), and in particular those with a high betweenness, tend to be highly constrained (Hahn et al. 2004; Vitkup et al. 2006; Lu et al. 2007; Greenberg et al. 2008; Hudson and Conant 2011; Montanucci et al. 2011), and enzymes that are connected in the metabolic network exhibit similar rates of evolution (Wang and Lercher 2011). In addition, enzymes carrying more intense metabolic fluxes tend to exhibit slower rates of evolution (Vitkup et al. 2006). On the contrary, in gene regulatory networks the most central transcription factors (i.e., those regulating a high number of other transcription factors, those regulated by a high number of transcription factors, or those sharing targets with a high number of other transcription factors) tend to evolve faster (Jovelin and Phillips 2009; Wang et al. 2010; Coulombe-Huntington and Xia 2012).

Proteins' relative hierarchical positions in molecular pathways also have an effect on their rates of evolution. However, analysis of biosynthetic, gene regulatory, and signal transduction pathways has provided contrasting results. In biosynthetic pathways, genes acting at the upstream part are often more selectively constrained than those acting at the downstream part, a pattern that has been observed, so far, in the plant anthocyanin (Rausher et al. 1999; Lu and Rausher 2003; Rausher et al. 2008), isoprene (Sharkey et al. 2005), terpenoid (Ramsay et al. 2009) and carotenoid pathways (Livingstone and Anderson 2009; Clotault et al. 2012), the *Bombyx* melanin pathway (Yu et al. 2011), and the mammalian dopamine pathway (Ma et al. 2010)—in the primate N-glycosylation pathway, however, the opposite trend has been described (Montanucci et al. 2011). This polarity in the levels of selective constraint along the upstream/downstream axis of biosynthetic pathways could be explained by two non-exclusive scenarios. First, biosynthetic pathways often have multiple bifurcations that lead to the biosynthesis of different end products. As a result of this branching topology, the most upstream enzyme (that catalyzing the first step of the pathway) is involved in the biosynthesis of the full set of biochemical compounds that are produced by the pathway; however, as we move towards the downstream part of the pathway, enzymes are required for the synthesis of a progressively smaller subset of these compounds. Mutations affecting upstream genes would hence be expected to have higher pleiotropic effects, thus resulting in lower rates of evolution for upstream proteins (Rausher et al. 1999; Ramsay et al. 2009). Second, simulation analyses of hypothetical biosynthetic pathways, as well as experimental manipulation of the *A. thaliana* glucosinolate pathway, have shown that upstream enzymes tend to exert a higher influence over metabolic fluxes along pathways (i.e., they exhibit higher control coefficients; Kacser and Burns 1973), and thus may be subject to stronger selective pressures (Wright and Rausher 2010; Olson-Manning et al. 2013; Rausher 2013).

Less well understood is how the hierarchical position of transcription factors in gene regulatory networks affects their rate of evolution, but at least two lines of evidence suggest that levels of selective constraint may follow a similar distribution to that observed in biosynthetic pathways. First, genes occupying highly hierarchical positions in these networks are more likely to be essential (Bhardwaj et al. 2010). Second, simulation analyses have suggested that genes occupying highly hierarchical positions (i.e., those exerting a higher degree of control over other genes, and/or those that are less regulated by other genes), are subject to stronger selective pressures (Rhone et al. 2011). Furthermore, transcription factors regulated by a high number of other transcription factors (thus occupying lowly hierarchical positions) tend to evolve fast (Wang et al. 2010), although those that regulate a high number of other transcription factors (thus occupying high hierarchies) also tend to evolve fast (Wang et al. 2010; Coulombe-Huntington and Xia 2012).

Signal transduction pathways include receptors that are capable, upon interaction with extracellular or intracellular stimuli, of unleashing a cascade of signals that end up activating the pathway effectors, which in turn mediate the relevant cellular responses. Surprisingly, in contrast with the patterns generally observed in biosynthetic pathways, in signalling pathways downstream genes tend to be the most selectively constrained. In the insulin/TOR pathway (Alvarez-Ponce et al. 2009, 2011, 2013a; Jovelin and Phillips 2011; Alvarez-Ponce et al. 2012; Wang et al. 2013), the yeast HOG pathway (Wu et al. 2010), and the vertebrate Toll-like receptor pathway (Song et al. 2012), proteins' rates of evolution significantly correlate with pathway position (computed as the number of steps required to transduce the signal from the receptors to the rest of pathway components), with downstream genes being progressively more selectively constrained. In addition, analysis of the human global signal transduction network provides similar results: proteins acting at the downstream part of the network tend to evolve faster than those occupying highly hierarchical positions, probably because the latter are more highly and broadly expressed, and more highly connected (Alvarez-Ponce 2012). Nevertheless, despite this general trend, not all particular pathways follow this polarity (Riley et al. 2003; Fitzpatrick and O'Halloran 2012), and the particularities of each pathway may ultimately determine the distribution of selective constraints. The different distributions of levels of selective constraint along biosynthetic and signal transduction pathways might reflect their different dynamics (see Alvarez-Ponce 2012).

Finally, beyond network centrality and upstream/ downstream position, other, less intuitive aspects of the position of proteins in molecular systems may have an effect on their rates of evolution. Owing to the particular structure and dynamics of each pathway, mutations affecting the kinetics of certain enzymes may have drastic effects on the overall pathway behaviour, whereas mutations affecting other enzymes may have little or no effect on the pathway function (and hence, on the associated phenotypes) (Kacser and Burns 1973). Enzymes exerting a high influence on the function of the pathway are expected to be subject to stronger selective pressures. Consistently, genes acting at bifurcation points of some pathways have been reported to evolve under positive selection (Flowers et al. 2007; Eanes 2011; Dall'Olio et al. 2012; Rausher 2013), and proteins with a greater influence on the dynamics of the pathways of which they are part tend to be more selectively constrained (Gutenkunst 2009).

7.10 Other Factors Influencing Proteins' Rates of Evolution

In this section, some other factors that have been shown or suggested to have an effect on proteins' rates of evolution are briefly mentioned.

Several works have illustrated that proteins performing different functions, or those involved in different biological processes, differ in their average rates of evolution (e.g., Greenberg et al. 2008). For instance, enzymes evolve slower than non-enzymatic proteins (Rocha and Danchin 2004; Greenberg et al. 2008; Hudson and Conant 2011), *Drosophila* proteins involved in protein synthesis evolve slower than those involved in the metabolism of xenobiotics (Greenberg et al. 2008), and *Arabidopsis* proteins involved in sugar metabolism evolve slower than those involved in transcription (Alvarez-Ponce and Fares 2012). Furthermore, multifunctional proteins tend to be highly constrained, and the number of functions in which proteins are involved negatively correlates with their rates of evolution (Wilson et al. 1977; Salathé et al. 2006; Podder et al. 2009).

Genes of different ages (i.e., originated at different evolutionary times) have been reported to evolve at different rates, with ancient (i.e., broadly distributed) genes being more conserved than novel (i.e., lineage-specific) genes (Domazet-Loso and Tautz 2003; Daubin and Ochman 2004; Albà and Castresana 2005; García-Vallvé et al. 2005; Luz et al. 2006). This observation was claimed to be an artefact due to the difficulty of detecting distant homologs of fast-evolving genes (Elhaik et al. 2006). However, subsequent simulation analyses and genomic analyses using the genomes of closely-related organisms confirmed that young genes indeed evolve faster than ancient onces (Albà and Castresana 2007; Toll-Riera et al. 2008; Wolf et al. 2009; Vishnoi et al. 2010; Toll-Riera et al. 2012).

Analyses in both *C. elegans* (Cutter and Ward 2005) and *D. melanogaster* (Davis et al. 2005) have shown that proteins acting early in development evolve slower than those acting in the adult. At least three explanations have been proposed. First, genes acting at early stages may be more important because later stages of development depend on successful execution of the previous ones (Riedl 1978; Arthur 1988, 1997). Second, theories of senescence predict relaxed selective pressures on genes acting in the later stages of life, when the reproductive capability decreases (Medawar 1952; Charlesworth 1994; Promislow and Tatar 1998; Partridge 2001). Third, certain developmental stages may be more sensitive to perturbation than others because of their particular architectures (Goodwin et al. 1993; Raff 1996).

Genes subject to tighter regulation seem to evolve slower than those that are less regulated. For instance, it has been shown that both the number of transcription factors (Xia et al. 2009; Wang et al. 2010) and the number of microRNAs (Cheng et al. 2009; Chen et al. 2011) by which a gene is

156

regulated negatively correlate with the rates of evolution of the encoded protein. Nevertheless, a recent study suggests that microRNAs may have a stronger effect (Chen et al. 2013).

Genes with a high codon usage bias tend to encode slow-evolving proteins (Sharp and Li 1987; Sharp 1991; Akashi 1994; Pál et al. 2001b). The reasons for this trend, however, remain poorly understood, and several explanations have been proposed. First, non-synonymous mutations at preferred codons can entail slightly deleterious changes to unpreferred codons, which may result in decreased d_N values for genes with high codon usage bias (Akashi 2001; Akashi 2003). Second, the most important amino acids in a protein may tend to be highly constrained and, at the same time, to be encoded by optimal codons (Akashi 1994). Third, genes that undergo a relaxation of purifying selection acting on non-synonymous sites may also experience relaxation at the level of codon bias (Comeron and Kreitman 1998). Fourth, high levels of codon bias often result in high levels of translation, and according to both the functional loss (Rocha and Danchin 2004) and the translational robustness (Drummond et al. 2005) hypotheses (developed above), highly translated proteins would tend to be highly constrained. Fifth, it has been proposed that fixation of adaptive nonsynonymous mutations would interfere with weak selection for codon usage at linked positions, thus resulting in fast-evolving proteins exhibiting low levels of codon bias (Betancourt and Presgraves 2002). Finally, codon bias strongly correlates with gene expression levels, raising the possibility that the correlation between codon bias and proteins' rates of evolution could be a by-product of the correlation of both variables with expression levels rather than a direct effect. Indeed, Ingvarsson (2007) detected no

157

association between d_N/d_S and codon bias once the effects of expression level and breadth were discounted.

As developed above, linked positions that are simultaneously under selection can interfere with each other, thus reducing the efficacy of both positive and negative selection (the so-called Hill-Robertson effect; Hill and Robertson 1966). This effect can be alleviated by recombination, which breaks up the linkage between loci. As a result, genes located at high-recombination genomic regions can experience increased purifying and positive selection. Therefore, proteins encoded by these genes can experience either increased or decreased rates of evolution, depending on whether advantageous or deleterious mutations are the most abundant, respectively. Consistently, among accelerated *Drosophila* genes, recombination rates positively correlate with proteins' rates of evolution (Betancourt and Presgraves 2002; Zhang and Parsch 2005; Larracuente et al. 2008), while the opposite trend is observed in non-accelerated genes (Marais et al. 2004; Haddrill et al. 2007; Larracuente et al. 2008). In yeasts, recombination rates negatively correlate with proteins' rates of evolution, although the trend disappears when the confounding effect of expression level is removed (Pál et al. 2001a). In addition to its effect on the efficacy of natural selection, recombination also may be mutagenic (Marais et al. 2001; Lercher and Hurst 2002; Filatov and Gerrard 2003; Hellmann et al. 2003) and might hence increase proteins' rates of evolution.

Eukaryotes are thought to have arisen about 2 billion years ago from a fusion event involving an archaebacterium and a eubacterium (Sagan 1967; Martin et al. 2001). As a result, eukaryotic genomes are chimeras containing genes of both archaebacterial and eubacterial ancestry (Gabaldón and Huynen 2003; Rivera and Lake 2004; Pisani et al. 2007;

Alvarez-Ponce et al. 2013b). Remarkably, eukaryotic genes contributed by both prokaryotic ancestors of Eukaryotes differ in a number of ways, e.g., performing different functions, acting at different subcellular compartments, having different probabilities of being essential, being expressed at different levels, and occupying different positions in the protein–protein interaction network (Rivera et al. 1998; Esser et al. 2004; Cotton and McInerney 2010; Alvarez-Ponce and McInerney 2011; Alvarez-Ponce et al. 2013b). In addition, genes of archaebacterial ancestry tend to be more selectively constrained than those of eubacterial ancestry (Alvarez-Ponce and McInerney 2011).

7.11 Correlation Does not Imply Causation: Looking for True Determinants of Proteins' Rates of Evolution in an Entanglement of Intercorrelated Factors

Identifying a correlation between a given factor and d_N/d_S is a key step towards establishing it as a determinant of rates of evolution. However, not all correlates of d_N/d_S are necessarily determinants of rates of evolution: some may simply be factors that correlate with other factors that, in turn, are determinants of rates of evolution. For instance, one can think of a hypothetical factor, termed X, that does not have an effect on proteins' rates of evolution, but that correlates with levels of gene expression; X will then probably correlate with rates of evolution, as a byproduct of its correlation with expression level (i.e., the correlation between d_N/d_S and X will be an indirect, induced correlation rather than the result of a true cause-effect relationship). Therefore, when a new correlate of d_N/d_S is discovered, it is crucial to rule out the possibility that this correlation is induced by potential confounding variables before the factor can be claimed to have a direct effect on evolutionary rates. Accounting for potential confounding factors can not only help discard spurious correlations (i.e., identifying false positives), but also can allow the manifestation of otherwise undetectable correlates of d_N/d_S (i.e., avoiding false negatives; e.g., Williams and Fares 2010; Lachowiec et al. 2013).

Given the great array of factors affecting proteins' rates of evolution, an integrated analysis of as many factors as possible can be essential. However, simultaneous analysis of several factors can be problematic, as the more factors are included in the analysis, the less genes exist with available information data for all variables. In any case, because gene expression seems to be the most important determinant of rates of evolution, discarding its effect is paramount.

The effect of potentially confounding factors can be minimized by applying a multitude of statistical techniques of varying complexity. When the factor that should be controlled is categorical (e.g., can take two or more pre-defined values), one can simply eliminate certain genes from the analysis (e.g., one can evaluate a certain correlation between a factor and d_N/d_S after eliminating genes with signatures of positive selection, in order to evaluate whether positive selection is affecting the trend; e.g., Alvarez-Ponce 2012), or bin the dataset into different categories (e.g., one can evaluate the correlation separately among singleton and duplicated genes, to discard duplicability as the driving factor; e.g., Alvarez-Ponce and Fares 2012). When the variables under consideration are quantitative, and particularly when multiple variables are to be handled simultaneously, multivariate analysis techniques are the methods of choice. Techniques of this kind that have been applied to understand the determinants of proteins' rates of evolution include partial correlation (Bloom and Adami

2003; Larracuente et al. 2008; Figure 7.3), multiple regression (Rocha and Danchin 2004; Xia et al. 2009), principal component analysis (Wolf et al. 2006), principal component regression analysis (Drummond et al. 2006; Yang and Gaut 2011), and structural equation modelling (Fraser et al. 2002; Wall et al. 2005; Ingvarsson 2007). Importantly, it is not clear which of these techniques is the most appropriate to study the determinants of evolutionary rates and, despite

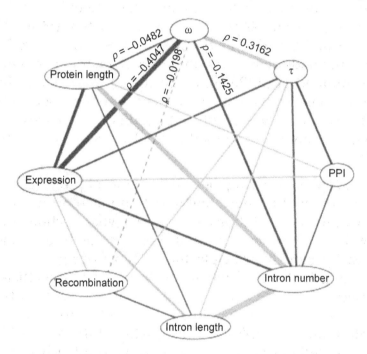

Figure 7.3. **Partial correlations among evolutionary rates (ω) and a number of genomic factors in** *Drosophila.* PPI, number of protein–protein interactions; τ, range of tissue expression. Positive correlations are represented in orange, and negative correlations are represented in blue. The width of the lines is proportional to the strength of the correlations. Figure from Larracuente et al. (2008), reproduced with permission from the authors and the publisher. Copyright 2007 Elsevier Ltd.

Color image of this figure appears in the color plate section at the end of the book.

their indubitable informative value, arguments can be proposed against all of them (Drummond et al. 2006; Kim and Yi 2007; Plotkin and Fraser 2007). First, these methods rely on a number of assumptions that are not always met. For instance, most methods assume linear, or at least monotonic dependences between variables, although not all associations between genomic factors are linear (Carmel and Koonin 2009). Second, although many of the correlates of d_N/d_S are also correlated to each other (Koonin and Wolf 2006; Ingvarsson 2007; Larracuente et al. 2008; Xia et al. 2009; Yang et al. 2009; Alvarez-Ponce and Fares 2012; Figure 7.3), some methods do not account for these inter-dependences (Drummond et al. 2006), or do it in an over-simplified manner. Finally, these methods may be confounded by the different accuracy with which the different variables have been measured (Drummond et al. 2006; Plotkin and Fraser 2007).

Beyond identifying the determinants of proteins' rates of evolution, it is interesting to determine to what extend they influence evolutionary rates, i.e., what fraction of the variability of proteins' rates of evolution can be explained by each factor? In principle, the influence of a given factor can be estimated as ρ^2, where ρ is the correlation coefficient between the factor and d_N/d_S. However, correlation coefficients are sensitive to a number of factors, such as confounding variables and the quality of the data, which complicate the estimation of the contribution of each factor to the variability of proteins' rates of evolution. Unfortunately, currently available large-scale datasets are subject to very high noise-to-signal ratios, which may result in decreased correlation coefficients. The quality of the measurements of the relevant variables can widely vary from one variable to another. For instance, if we consider the set of factors that affect the rates of evolution of *E. coli* proteins, for certain factors, such as proteins' lengths, we will have very precise measures (as the *E. coli* genome is very well annotated); on the contrary, available measures for other factors such as genes' levels of expression and number of protein–protein interactions may be much noisier, as they are the result, to a great extent, of the application of high-throughput techniques. This could result in an under-estimation of the relative effect of the latter, noisily-measured factors, on proteins' rates of evolution. Indeed, Plotkin and Fraser (2007) found that, once noise levels were equalized across 7 predictors of d_N/d_S in yeasts, they all seem to have a roughly equal contribution to the variability of proteins' rates of evolution.

7.12 Concluding Remarks

In summary, proteins' rates of evolution are known to be influenced by a plethora of factors, and with no doubt many more will continue to be discovered in the near future. The prevailing view is that patterns and levels of gene expression are the most important factors affecting evolutionary rates, and that the other factors have a relatively minor, albeit measurable, effect. Although this might well be true, further experimental and theoretical advances are warranted to gain further insight into the relative contributions of the different factors to the variability of proteins' rates of evolution, and into the mechanisms by which they affect evolutionary rates.

Acknowledgements

The author is grateful to Mavio A. Fares for support and helpful comments on the manuscript. The author is a *Juan de la Cierva* postdoctoral fellow from the Spanish Ministerio de Economía y Competitividad (JCI-2011-11089).

References

Agrafioti, I., J. Swire, J. Abbott, D. Huntley, S. Butcher and M.P. Stumpf. 2005. Comparative analysis of the Saccharomyces cerevisiae and Caenorhabditis elegans protein interaction networks. BMC Evol Biol 5: 23.

Akashi, H. 1994. Synonymous codon usage in Drosophila melanogaster: natural selection and translational accuracy. Genetics 136: 927–935.

Akashi, H. 2001. Gene expression and molecular evolution. Curr Opin Genet Dev 11: 660–666.

Akashi, H. 2003. Translational selection and yeast proteome evolution. Genetics 164: 1291–1303.

Albà, M.M. and J. Castresana. 2005. Inverse relationship between evolutionary rate and age of mammalian genes. Mol Biol Evol 22: 598–606.

Albà, M.M. and J. Castresana. 2007. On homology searches by protein Blast and the characterization of the age of genes. BMC Evol Biol. 7: 53.

Alvarez-Ponce, D. 2012. The relationship between the hierarchical position of proteins in the human signal transduction network and their rate of evolution. BMC Evol Biol 12: 192.

Alvarez-Ponce, D., M. Aguadé and J. Rozas. 2009. Network-level molecular evolutionary analysis of the insulin/TOR signal transduction pathway across 12 Drosophila genomes. Genome Res 19: 234–242.

Alvarez-Ponce, D., M. Aguadé and J. Rozas. 2011. Comparative Genomics of the Vertebrate Insulin/TOR Signal Transduction Pathway: A Network-Level Analysis of Selective Pressures. Genome Biol Evol 3: 87–101.

Alvarez-Ponce, D. and M.A. Fares. 2012. Evolutionary Rate and Duplicability in the Arabidopsis thaliana Protein-Protein Interaction Network. Genome Biol Evol 4: 1263–1274.

Alvarez-Ponce, D., S. Guirao-Rico, D.J. Orengo, C. Segarra, J. Rozas and M. Aguadé. 2012. Molecular Population Genetics of the Insulin/TOR Signal Transduction Pathway: A Network-Level Analysis in Drosophila melanogaster. Mol Biol Evol 29: 123–132.

Alvarez-Ponce, D., M. Aguade and J. Rozas. 2013a. Comment on "The Molecular Evolutionary Patterns of the Insulin/FOXO Signaling Pathway". Evol Bioinform Online 9: 229–234.

Alvarez-Ponce, D., P. Lopez, E. Bapteste and J.O. McInerney. 2013b. Gene similarity networks provide tools for understanding eukaryote origins and evolution. Proc Natl Acad Sci U S A 110: E1594–1603.

Alvarez-Ponce, D. and J.O. McInerney. 2011. The human genome retains relics of its prokaryotic ancestry: human genes of archaebacterial and eubacterial origin exhibit remarkable differences. Genome Biol Evol 3: 782–790.

Arthur, W. 1988. A Theory of the Evolution of Development. John Wiley & Sons, Chichester.

Arthur, W. 1997. The Origin of Animal Body Plants: a Study in Evolutionary Developmental Biology. Cambridge University Press, New York.

Bader, J.S., A. Chaudhuri, J.M. Rothberg and J. Chant. 2004. Gaining confidence in high-throughput protein interaction networks. Nat Biotechnol 22: 78–85.

Baumann, P., L. Baumann and M.A. Clark. 1996. Levels of Buchnera aphidicola chaperonin GroEL during growth of the aphid Schizaphis graminum. Current Microbiology 32: 279–285.

Betancourt, A.J. and D.C. Presgraves. 2002. Linkage limits the power of natural selection in Drosophila. Proc Natl Acad Sci USA 99: 13616–13620.

Bhardwaj, N., P.M. Kim and M.B. Gerstein. 2010. Rewiring of transcriptional regulatory networks: hierarchy, rather than connectivity, better reflects the importance of regulators. Sci Signal 3: ra79.

Bloom, J.D. and C. Adami. 2003. Apparent dependence of protein evolutionary rate on number of interactions is linked to biases in protein-protein interactions data sets. BMC Evol Biol 3: 21.

Bloom, J.D. and C. Adami. 2004. Evolutionary rate depends on number of protein-protein interactions independently of gene expression level: response. BMC Evol Biol 4: 14.

Bloom, J.D., D.A. Drummond, F.H. Arnold and C.O. Wilke. 2006a. Structural determinants of the rate of protein evolution in yeast. Mol Biol Evol 23: 1751–1761.

Bloom, J.D., S.T. Labthavikul, C.R. Otey and F.H. Arnold. 2006b. Protein stability promotes evolvability. Proc Natl Acad Sci U S A 103: 5869–5874.

Bogumil, D. and T. Dagan. 2010. Chaperonin-dependent accelerated substitution rates in prokaryotes. Genome Biol Evol 2: 602–608.

Bogumil, D. and T. Dagan. 2012. Cumulative impact of chaperone-mediated folding on genome evolution. Biochemistry 51: 9941–9953.

Bogumil, D., G. Landan, J. Ilhan and T. Dagan. 2012. Chaperones divide yeast proteins into classes of expression level and evolutionary rate. Genome Biol Evol 4: 618–625.

Brown, C.J., S. Takayama, A.M. Campen, P. Vise, T.W. Marshall, C.J. Oldfield, C.J. Williams and A.K. Dunker. 2002. Evolutionary rate heterogeneity in proteins with long disordered regions. J Mol Evol 55: 104–110.

Bustamante, C.D., J.P. Townsend and D.L. Hartl. 2000. Solvent accessibility and purifying selection within proteins of *Escherichia coli* and Salmonella enterica. Mol Biol Evol 17: 301–308.

Carmel, L. and E.V. Koonin. 2009. A universal nonmonotonic relationship between gene compactness and expression levels in multicellular eukaryotes. Genome Biol Evol 1: 382–390.

Casals, F., M. Sikora, H. Laayouni, L. Montanucci, A. Muntasell, R. Lazarus, F. Calafell, P. Awadalla, M.G. Netea and J. Bertranpetit. 2011. Genetic adaptation of the antibacterial human innate immunity network. BMC Evol Biol 11: 202.

Castillo-Davis, C.I. and D.L. Hartl. 2003. Conservation, relocation and duplication in genome evolution. Trends Genet 19: 593–597.

Chain, F.J. and B.J. Evans. 2006. Multiple mechanisms promote the retained expression of gene duplicates in the tetraploid frog Xenopus laevis. PLoS Genet 2: e56.

Chang, T.Y. and B.Y. Liao. 2013. Flagellated algae protein evolution suggests the prevalence of lineage-specific rules governing evolutionary rates of eukaryotic proteins. Genome Biol Evol.

Chatr-Aryamontri, A., B.J. Breitkreutz, S. Heinicke, et al. 2013. The BioGRID interaction database: 2013 update. Nucleic Acids Res 41: D816–823.

Chen, S.C., T.J. Chuang and W.H. Li. 2011. The relationships among microRNA regulation, intrinsically disordered regions, and other indicators of protein evolutionary rate. Mol Biol Evol 28: 2513–2520.

Chen, Y. and D. Xu. 2005. Understanding protein dispensability through machine-learning analysis of high-throughput data. Bioinformatics 21: 575–581.

Chen, Y.C., J.H. Cheng, Z.T. Tsai, H.K. Tsai and T.J. Chuang. 2013. The Impact of Trans-Regulation on the Evolutionary Rates of Metazoan Proteins. Nucleic Acids Res 41(13): 6371–6380.

Cheng, C., N. Bhardwaj and M. Gerstein. 2009. The relationship between the evolution of microRNA targets and the length of their UTRs. BMC Genomics 10: 431.

Cherry, J.L. 2010. Expression level, evolutionary rate, and the cost of expression. Genome Biol Evol 2: 757–769.

Clark, N.L., E. Alani and C.F. Aquadro. 2012. Evolutionary rate covariation reveals shared functionality and coexpression of genes. Genome Res 22: 714–720.

Clark, N.L. and C.F. Aquadro. 2010. A novel method to detect proteins evolving at correlated rates: identifying new functional relationships between coevolving proteins. Mol Biol Evol 27: 1152–1161.

Clotault, J., D. Peltier, V. Soufflet-Freslon, M. Briard and E. Geoffriau. 2012. Differential selection on carotenoid biosynthesis genes as a function of gene position in the metabolic pathway: a study on the carrot and dicots. PLoS One 7: e38724.

Comeron, J.M. and M. Kreitman. 1998. The correlation between synonymous and nonsynonymous substitutions in Drosophila: mutation, selection or relaxed constraints? Genetics 150: 767–775.

Comeron, J.M. and M. Kreitman. 2000. The correlation between intron length and recombination in drosophila. Dynamic equilibrium between mutational and selective forces. Genetics 156: 1175–1190.

Comeron, J.M. and M. Kreitman. 2002. Population, evolutionary and genomic consequences of interference selection. Genetics 161: 389–410.

Conant, G.C. 2009. Neutral evolution on mammalian protein surfaces. Trends Genet 25: 377–381.

Conant, G.C. and A. Wagner. 2003. Asymmetric sequence divergence of duplicate genes. Genome Res 13: 2052–2058.

Cork, J.M. and M.D. Purugganan. 2004. The evolution of molecular genetic pathways and networks. Bioessays 26: 479–484.

Cotton, J.A. and J.O. McInerney. 2010. Eukaryotic genes of archaebacterial origin are more important than the more numerous eubacterial genes, irrespective of function. Proc Natl Acad Sci U S A 107: 17252–17255.

Coulombe-Huntington, J. and Y. Xia. 2012. Regulatory network structure as a dominant determinant of transcription factor evolutionary rate. PLoS Comput Biol 8: e1002734.

Cui, Q., E.O. Purisima and E. Wang. 2009. Protein evolution on a human signaling network. BMC Syst Biol 3: 21.

Cusack, B.P. and K.H. Wolfe. 2007. Not born equal: increased rate asymmetry in relocated and retrotransposed rodent gene duplicates. Mol Biol Evol 24: 679–686.

Cutter, A.D. and S. Ward. 2005. Sexual and temporal dynamics of molecular evolution in C. elegans development. Mol Biol Evol 22: 178–188.

Dall'Olio, G.M., H. Laayouni, P. Luisi, M. Sikora, L. Montanucci and J. Bertranpetit. 2012. Distribution of events of positive selection and population differentiation in a metabolic pathway: the case of asparagine N-glycosylation. BMC Evol Biol 12: 98.

Daubin, V. and H. Ochman. 2004. Bacterial genomes as new gene homes: the genealogy of ORFans in E. coli. Genome Res 14: 1036–1042.

Davids, W. and Z. Zhang. 2008. The impact of horizontal gene transfer in shaping operons and protein interaction networks—direct evidence of preferential attachment. BMC Evol Biol 8: 23.

Davis, J.C., O. Brandman and D.A. Petrov. 2005. Protein evolution in the context of Drosophila development. J Mol Evol 60: 774–785.

Davis, J.C. and D.A. Petrov. 2004. Preferential duplication of conserved proteins in eukaryotic genomes. PLoS Biol 2: E55.

Dean, M.D., J.M. Good and M.W. Nachman. 2008. Adaptive evolution of proteins secreted during sperm maturation: an analysis of the mouse epididymal transcriptome. Mol Biol Evol 25: 383–392.

Deeds, E.J., O. Ashenberg and E.I. Shakhnovich. 2006. A simple physical model for scaling in protein-protein interaction networks. Proc Natl Acad Sci U S A 103: 311–316.

Dickerson, R.E. 1971. The structures of cytochrome c and the rates of molecular evolution. J Mol Evol 1: 26–45.

Doherty, A., D. Alvarez-Ponce and J.O. McInerney. 2012. Increased Genome Sampling Reveals a Dynamic Relationship between Gene Duplicability and the Structure of the Primate Protein-Protein Interaction Network. Mol Biol Evol 29: 3563–3573.

167

Domazet-Loso, T. and D. Tautz. 2003. An evolutionary analysis of orphan genes in Drosophila. Genome Res 13: 2213–2219.

Dötsch, A., F. Klawonn, M. Jarek, M. Scharfe, H. Blocker and S. Haussler. 2010. Evolutionary conservation of essential and highly expressed genes in Pseudomonas aeruginosa. BMC Genomics 11: 234.

Drummond, D.A., J.D. Bloom, C. Adami, C.O. Wilke and F.H. Arnold. 2005. Why highly expressed proteins evolve slowly. Proc Natl Acad Sci U S A 102: 14338–14343.

Drummond, D.A., A. Raval and C.O. Wilke. 2006. A single determinant dominates the rate of yeast protein evolution. Mol Biol Evol 23: 327–337.

Drummond, D.A. and C.O. Wilke. 2008. Mistranslation-induced protein misfolding as a dominant constraint on coding-sequence evolution. Cell 134: 341–352.

Duret, L. and D. Mouchiroud. 1999. Expression pattern and, surprisingly, gene length shape codon usage in Caenorhabditis, Drosophila, and Arabidopsis. Proc Natl Acad Sci U S A 96: 4482–4487.

Duret, L. and D. Mouchiroud. 2000. Determinants of substitution rates in mammalian genes: expression pattern affects selection intensity but not mutation rate. Mol Biol Evol 17: 68–74.

Eanes, W.F. 1999. Analysis of selection on enzyme polymorphisms. Rev Ecol Syst 30: 301–326.

Eanes, W.F. 2011. Molecular population genetics and selection in the glycolytic pathway. J Exp Biol 214: 165–171.

Elhaik, E., N. Sabath and D. Graur. 2006. The "inverse relationship between evolutionary rate and age of mammalian genes" is an artifact of increased genetic distance with rate of evolution and time of divergence. Mol Biol Evol 23: 1–3.

Esser, C., N. Ahmadinejad, C. Wiegand, et al. 2004. A genome phylogeny for mitochondria among alpha-proteobacteria and a predominantly eubacterial ancestry of yeast nuclear genes. Mol Biol Evol 21: 1643–1660.

Fares, M.A., M.X. Ruiz-Gonzalez and J.P. Labrador. 2011. Protein coadaptation and the design of novel approaches to identify protein-protein interactions. IUBMB Life 63: 264–271.

Fares, M.A., M.X. Ruiz-Gonzalez, A. Moya, S.F. Elena and E. Barrio. 2002. Endosymbiotic bacteria: groEL buffers against deleterious mutations. Nature 417: 398.

Filatov, D.A. and D.T. Gerrard. 2003. High mutation rates in human and ape pseudoautosomal genes. Gene 317: 67–77.

Fitzpatrick, D.A. and D.M. O'Halloran. 2012. Investigating the Relationship between Topology and Evolution in a Dynamic Nematode Odor Genetic Network. Int J Evol Biol 548081.

Flowers, J.M., E. Sezgin, S. Kumagai, D.D. Duvernell, L.M. Matzkin, P.S. Schmidt and W.F. Eanes. 2007. Adaptive evolution of metabolic pathways in Drosophila. Mol Biol Evol 24: 1347–1354.

Franzosa, E.A. and Y. Xia. 2009. Structural determinants of protein evolution are context-sensitive at the residue level. Mol Biol Evol 26: 2387–2395.

Fraser, H.B. 2005. Modularity and evolutionary constraint on proteins. Nat Genet 37: 351–352.

Fraser, H.B. and A.E. Hirsh. 2004. Evolutionary rate depends on number of protein-protein interactions independently of gene expression level. BMC Evol Biol 4: 13.

Fraser, H.B., A.E. Hirsh, L.M. Steinmetz, C. Scharfe and M.W. Feldman. 2002. Evolutionary rate in the protein interaction network. Science 296: 750–752.

Fraser, H.B., D.P. Wall and A.E. Hirsh. 2003. A simple dependence between protein evolution rate and the number of protein-protein interactions. BMC Evol Biol 3: 11.

Freeman, L.C. 1977. A set of measures of centrality based on betweenness. Sociometry 35–41.

Fryxell, K.J. 1996. The coevolution of gene family trees. Trends Genet 12: 364–369.

Gabaldón, T. and M.A. Huynen. 2003. Reconstruction of the proto-mitochondrial metabolism. Science 301: 609.

García-Vallvé, S., A. Alonso and I.G. Bravo. 2005. Papillomaviruses: different genes have different histories. Trends Microbiol 13: 514–521.

Geiler-Samerotte, K.A., M.F. Dion, B.A. Budnik, S.M. Wang, D.L. Hartl and D.A. Drummond. 2010. Misfolded proteins impose a dosage-dependent fitness cost and trigger a cytosolic unfolded protein response in yeast. Proc Natl Acad Sci U S A 108: 680–685.

Goldman, N., J.L. Thorne and D.T. Jones. 1998. Assessing the impact of secondary structure and solvent accessibility on protein evolution. Genetics 149: 445–458.

Goodwin, B.C., S. Kauffman and J.D. Murray. 1993. Is morphogenesis an intrinsically robust process? J Theor Biol 163: 135–144.

Gout, J.F., D. Kahn and L. Duret. 2010. The relationship among gene expression, the evolution of gene dosage, and the rate of protein evolution. PLoS Genet 6: e1000944.

Greenberg, A.J., S.R. Stockwell and A.G. Clark. 2008. Evolutionary constraint and adaptation in the metabolic network of Drosophila. Mol Biol Evol 25: 2537–2546.

Gutenkunst, R.N. 2009. Proteins with greater influence on network dynamics evolve more slowly but are not more essential. arXiv preprint arXiv:0909.2889.

Haddrill, P.R., D.L. Halligan, D. Tomaras and B. Charlesworth. 2007. Reduced efficacy of selection in regions of the Drosophila genome that lack crossing over. Genome Biol 8: R18.

Hahn, M.W. 2009. Distinguishing among evolutionary models for the maintenance of gene duplicates. J Hered 100: 605–617.

169

Hahn, M.W., G.C. Conant and A. Wagner. 2004. Molecular evolution in large genetic networks: does connectivity equal constraint? J Mol Evol 58: 203–211.

Hahn, M.W. and A.D. Kern. 2005. Comparative genomics of centrality and essentiality in three eukaryotic protein-interaction networks. Mol Biol Evol 22: 803–806.

Han, M.V., J.P. Demuth, C.L. McGrath, C. Casola and M.W. Hahn. 2009. Adaptive evolution of young gene duplicates in mammals. Genome Res 19: 859–867.

He, X. and J. Zhang. 2005. Rapid subfunctionalization accompanied by prolonged and substantial neofunctionalization in duplicate gene evolution. Genetics 169: 1157–1164.

Heger, A., C.P. Ponting and I. Holmes. 2009. Accurate estimation of gene evolutionary rates using XRATE, with an application to transmembrane proteins. Mol Biol Evol 26: 1715–1721.

Hegyi, H. and P. Bork. 1997. On the classification and evolution of protein modules. J Protein Chem 16: 545–551.

Hellmann, I., I. Ebersberger, S.E. Ptak, S. Paabo and M. Przeworski. 2003. A neutral explanation for the correlation of diversity with recombination rates in humans. Am J Hum Genet 72: 1527–1535.

Henderson, B., M.A. Fares and P.A. Lund. 2013. Chaperonin 60: a paradoxical, evolutionarily conserved protein family with multiple moonlighting functions. Biol Rev Camb Philos Soc (in press).

Hershberg, R. and D.A. Petrov. 2008. Selection on codon bias. Annu Rev Genet 42: 287–299.

Hill, W.G. and A. Robertson. 1966. The effect of linkage on limits to artificial selection. Genet Res 8: 269–294.

Hirsh, A.E. and H.B. Fraser. 2001. Protein dispensability and rate of evolution. Nature 411: 1046–1049.

Hudson, C.M. and G.C. Conant. 2011. Expression level, cellular compartment and metabolic network position all influence the average selective constraint on mammalian enzymes. BMC Evol Biol 11: 89.

Hurst, L.D. and N.G. Smith. 1999. Do essential genes evolve slowly? Curr Biol 9: 747–750.

Ingram, V.M. 1961. Gene evolution and the haemoglobins. Nature 189: 704–708.

Ingvarsson, P.K. 2007. Gene expression and protein length influence codon usage and rates of sequence evolution in Populus tremula. Mol Biol Evol 24: 836–844.

Innan, H. and F. Kondrashov. 2010. The evolution of gene duplications: classifying and distinguishing between models. Nat Rev Genet 11: 97–108.

Invergo, B.M., L. Montanucci, H. Laayouni and J. Bertranpetit. 2013. A system-level, molecular evolutionary analysis of mammalian phototransduction. BMC Evol Biol 13: 52.

Jordan, I.K., I.B. Rogozin, Y.I. Wolf and E.V. Koonin. 2002. Essential genes are more evolutionarily conserved than are nonessential genes in bacteria. Genome Res 12: 962–968.

Jordan, I.K., Y.I. Wolf and E.V. Koonin. 2003. No simple dependence between protein evolution rate and the number of protein-protein interactions: only the most prolific interactors tend to evolve slowly. BMC Evol Biol 3: 1.

Jordan, I.K., Y.I. Wolf and E.V. Koonin. 2004. Duplicated genes evolve slower than singletons despite the initial rate increase. BMC Evol Biol 4: 22.

Jovelin, R. and P.C. Phillips. 2009. Evolutionary rates and centrality in the yeast gene regulatory network. Genome Biol 10: R35.

Jovelin, R. and P.C. Phillips. 2011. Expression Level Drives the Pattern of Selective Constraints along the Insulin/Tor Signal Transduction Pathway in Caenorhabditis. Genome Biol Evol 3: 715–722.

Julenius, K. and A.G. Pedersen. 2006. Protein evolution is faster outside the cell. Mol Biol Evol 23: 2039–2048.

Jun, J., P. Ryvkin, E. Hemphill and C. Nelson. 2009. Duplication mechanism and disruptions in flanking regions determine the fate of Mammalian gene duplicates. J Comput Biol 16: 1253–1266.

Kacser. H. and J.A. Burns. 1973. The control of flux. Symp Soc Exp Biol 27: 65–104.

Kellis, M., B.W. Birren and E.S. Lander. 2004. Proof and evolutionary analysis of ancient genome duplication in the yeast Saccharomyces cerevisiae. Nature 428: 617–624.

Kelly, W.P. and M.P. Stumpf. 2012. Assessing coverage of protein interaction data using capture-recapture models. Bull Math Biol 74: 356–374.

Kim, P.M., J.O. Korbel and M.B. Gerstein. 2007. Positive selection at the protein network periphery: evaluation in terms of structural constraints and cellular context. Proc Natl Acad Sci U S A 104: 20274–20279.

Kim, S.H. and S.V. Yi. 2007. Understanding relationship between sequence and functional evolution in yeast proteins. Genetica 131: 151–156.

Kimura, M. 1968. Evolutionary rate at the molecular level. Nature 217: 624–626.

Kimura, M. 1983. The Neutral Theory of Molecular Evolution. Cambridge University Press, Cambridge.

Kimura, M. and T. Ohta. 1974. On some principles governing molecular evolution. Proceedings of the National Academy of Sciences 71: 2848–2852.

Kisters-Woike, B., C. Vangierdegom and B. Müller-Hill. 2000. On the conservation of protein sequences in evolution. Trends Biochem Sci 25: 419–421.

Kondrashov, F.A., I.B. Rogozin, Y.I. Wolf and E.V. Koonin. 2002. Selection in the evolution of gene duplications. Genome Biol 3:RESEARCH0008.

Koonin, E.V. and Y.I. Wolf. 2006. Evolutionary systems biology: links between gene evolution and function. Curr Opin Biotechnol 17: 481–487.

Krylov, D.M., Y.I. Wolf, I.B. Rogozin and E.V. Koonin. 2003. Gene loss, protein sequence divergence, gene dispensability, expression level, and interactivity are correlated in eukaryotic evolution. Genome Res 13: 2229–2235.

Kuma, K., N. Iwabe and T. Miyata. 1995. Functional constraints against variations on molecules from the tissue level: slowly evolving brain-specific genes demonstrated by protein kinase and immunoglobulin supergene families. Mol Biol Evol 12: 123–130.

Kussell, E. 2005. The designability hypothesis and protein evolution. Protein Pept Lett 12: 111–116.

Lachowiec, J., T. Lemus, J.H. Thomas, P.J. Murphy, J.L. Nemhauser and C. Queitsch. 2013. The protein chaperone HSP90 can facilitate the divergence of gene duplicates. Genetics 193: 1269–1277.

Larracuente, A.M., T.B. Sackton, A.J. Greenberg, A. Wong, N.D. Singh, D. Sturgill, Y. Zhang, B. Oliver and A.G. Clark. 2008. Evolution of protein-coding genes in Drosophila. Trends Genet 24: 114–123.

Lavagnino, N., F. Serra, L. Arbiza, H. Dopazo and E. Hasson. 2012. Evolutionary Genomics of Genes Involved in Olfactory Behavior in the Drosophila melanogaster Species Group. Evol Bioinform Online 8: 89–104.

Lemos, B., B.R. Bettencourt, C.D. Meiklejohn and D.L. Hartl. 2005. Evolution of proteins and gene expression levels are coupled in Drosophila and are independently associated with mRNA abundance, protein length, and number of protein-protein interactions. Mol Biol Evol 22: 1345–1354.

Lercher, M.J. and L.D. Hurst. 2002. Human SNP variability and mutation rate are higher in regions of high recombination. Trends Genet 18: 337–340.

Levy, E.D., S. De and S.A. Teichmann. 2012. Cellular crowding imposes global constraints on the chemistry and evolution of proteomes. Proc Natl Acad Sci U S A 109: 20461–20466.

Li, H., R. Helling, C. Tang and N. Wingreen. 1996. Emergence of preferred structures in a simple model of protein folding. Science 273: 666–669.

Li, W.H., C.I. Wu and C.C. Luo. 1985. A new method for estimating synonymous and nonsynonymous rates of nucleotide substitution considering the relative likelihood of nucleotide and codon changes. Mol Biol Evol 2: 150–174.

Liao, B.Y., N.M. Scott and J. Zhang. 2006. Impacts of gene essentiality, expression pattern, and gene compactness on the evolutionary rate of mammalian proteins. Mol Biol Evol 23: 2072–2080.

Liao, B.Y., M.P. Weng and J. Zhang. 2010. Impact of extracellularity on the evolutionary rate of mammalian proteins. Genome Biol Evol 2: 39–43.

Lin, Y.S., W.L. Hsu, J.K. Hwang and W.H. Li. 2007. Proportion of solvent-exposed amino acids in a protein and rate of protein evolution. Mol Biol Evol 24: 1005–1011.

Liu, J., Y. Zhang, X. Lei and Z. Zhang. 2008. Natural selection of protein structural and functional properties: a single nucleotide polymorphism perspective. Genome Biol 9: R69.

Livingstone, K. and S. Anderson. 2009. Patterns of variation in the evolution of carotenoid biosynthetic pathway enzymes of higher plants. J Hered. 100: 754–761.

Lovell, S.C. and D.L. Robertson. 2010. An integrated view of molecular coevolution in protein-protein interactions. Mol Biol Evol 27: 2567–2575.

Lu, C., Z. Zhang, L. Leach, M.J. Kearsey and Z.W. Luo. 2007. Impacts of yeast metabolic network structure on enzyme evolution. Genome Biol 8: 407.

Lu, Y. and M.D. Rausher. 2003. Evolutionary rate variation in anthocyanin pathway genes. Mol Biol Evol 20: 1844–1853.

Luz, H., E. Staub and M. Vingron. 2006. About the interrelation of evolutionary rate and protein age. Genome Inform 17: 240–250.

Lynch, M. and J.S. Conery. 2000. The evolutionary fate and consequences of duplicate genes. Science 290: 1151–1155.

Lynch, M. and V. Katju. 2004. The altered evolutionary trajectories of gene duplicates. Trends Genet 20: 544–549.

Ma, X., Z. Wang and X. Zhang. 2010. Evolution of dopamine-related systems: biosynthesis, degradation and receptors. J Mol Evol 71: 374–384.

Maisnier-Patin, S., J.R. Roth, A. Fredriksson, T. Nystrom, O.G. Berg and D.I. Andersson. 2005. Genomic buffering mitigates the effects of deleterious mutations in bacteria. Nat Genet 37: 1376–1379.

Marais, G., T. Domazet-Loso, D. Tautz and B. Charlesworth. 2004. Correlated evolution of synonymous and nonsynonymous sites in Drosophila. J Mol Evol 59: 771–779.

Marais, G. and L. Duret. 2001. Synonymous codon usage, accuracy of translation, and gene length in Caenorhabditis elegans. J Mol Evol 52: 275–280.

Marais, G., D. Mouchiroud and L. Duret. 2001. Does recombination improve selection on codon usage? Lessons from nematode and fly complete genomes. Proc Natl Acad Sci U S A 98: 5688–5692.

Martin, W., M. Hoffmeister, C. Rotte and K. Henze. 2001. An overview of endosymbiotic models for the origins of eukaryotes, their ATP-producing organelles (mitochondria and hydrogenosomes), and their heterotrophic lifestyle. Biological Chemistry 382: 1521–1539.

Mintseris, J. and Z. Weng. 2005. Structure, function, and evolution of transient and obligate protein-protein interactions. Proc Natl Acad Sci U S A 102: 10930–10935.

Montanucci, L., H. Laayouni, G.M. Dall'Olio and J. Bertranpetit. 2011. Molecular evolution and network-level analysis of the N-glycosylation metabolic pathway across primates. Mol Biol Evol 28: 813–823.

Nembaware, V., K. Crum, J. Kelso and C. Seoighe. 2002. Impact of the presence of paralogs on sequence divergence in a set of mouse-human orthologs. Genome Res 12: 1370–1376.

Ohno, S. 1970. Evolution by Gene Duplication. Springer-Verlag, Berlin.

Olson-Manning, C.F., C.R. Lee, M.D. Rausher and T. Mitchell-Olds. 2013. Evolution of flux control in the glucosinolate pathway in Arabidopsis thaliana. Mol Biol Evol 30: 14–23.

173

Paape, T., T. Bataillon, P. Zhou, J.Y.K. T, R. Briskine, N.D. Young and P. Tiffin. 2013. Selection, genome-wide fitness effects and evolutionary rates in the model legume Medicago truncatula. Mol Ecol 22: 3525–3538.

Pagán, I., E.C. Holmes and E. Simon-Loriere. 2012. Level of gene expression is a major determinant of protein evolution in the viral order Mononegavirales. J Virol 86: 5253–5263.

Pál, C., B. Papp and L.D. Hurst. 2001a. Does the recombination rate affect the efficiency of purifying selection? The yeast genome provides a partial answer. Mol Biol Evol 18: 2323–2326.

Pál, C., B. Papp and L.D. Hurst. 2001b. Highly expressed genes in yeast evolve slowly. Genetics 158: 927–931.

Pál, C., B. Papp and L.D. Hurst. 2003. Genomic function: Rate of evolution and gene dispensability. Nature 421: 496–497; discussion 497–498.

Pál, C., B. Papp and M.J. Lercher. 2006. An integrated view of protein evolution. Nat Rev Genet 7: 337–348.

Panchin, A.Y., M.S. Gelfand, V.E. Ramensky and I.I. Artamonova. 2010. Asymmetric and non-uniform evolution of recently duplicated human genes. Biol Direct 5: 54.

Park, C., X. Chen, J.R. Yang and J. Zhang. 2013. Differential requirements for mRNA folding partially explain why highly expressed proteins evolve slowly. Proc Natl Acad Sci U S A 110: E678–686.

Pegueroles, C., S. Laurie and M.M. Albà. 2013. Accelerated Evolution after Gene Duplication: A Time-Dependent Process Affecting Just One Copy. Mol Biol Evol 30(8): 1830–1842.

Pisani, D., J.A. Cotton and J.O. McInerney. 2007. Supertrees disentangle the chimerical origin of eukaryotic genomes. Mol Biol Evol 24: 1752–1760.

Plotkin, J.B. and H.B. Fraser. 2007. Assessing the determinants of evolutionary rates in the presence of noise. Mol Biol Evol 24: 1113–1121.

Podder, S., P. Mukhopadhyay and T.C. Ghosh. 2009. Multifunctionality dominantly determines the rate of human housekeeping and tissue specific interacting protein evolution. Gene 439: 11–16.

Popescu, C.E., T. Borza, J.P. Bielawski and R.W. Lee. 2006. Evolutionary rates and expression level in Chlamydomonas. Genetics 172: 1567–1576.

Queitsch, C., T.A. Sangster and S. Lindquist. 2002. Hsp90 as a capacitor of phenotypic variation. Nature 417: 618–624.

Raff, R.A. 1996. The Shape of Life: Genes, Development, and the Evolution of Animal Form. University of Chicago Press, Chicago.

Ramsay, H., L.H. Rieseberg and K. Ritland. 2009. The correlation of evolutionary rate with pathway position in plant terpenoid biosynthesis. Mol Biol Evol 26: 1045–1053.

Rausher, M.D. 2013. The evolution of genes in branched metabolic pathways. Evolution 67: 34–48.

Rausher, M.D., Y. Lu and K. Meyer. 2008. Variation in constraint versus positive selection as an explanation for evolutionary rate variation among anthocyanin genes. J Mol Evol 67: 137–144.

Rausher, M.D., R.E. Miller and P. Tiffin. 1999. Patterns of evolutionary rate variation among genes of the anthocyanin biosynthetic pathway. Mol Biol Evol 16: 266–274.

Ravasz, E., A.L. Somera, D.A. Mongru, Z.N. Oltvai and A.L. Barabasi. 2002. Hierarchical organization of modularity in metabolic networks. Science 297: 1551–1555.

Renaut, S., C.J. Grassa, B.T. Moyers, N.C. Kane and L.H. Rieseberg. 2012. The Population Genomics of Sunflowers and Genomic Determinants of Protein Evolution Revealed by RNAseq. Biology 1: 575–596.

Rhone, B., J.T. Brandenburg and F. Austerlitz. 2011. Impact of selection on genes involved in regulatory network: a modelling study. J Evol Biol 24: 2087–2098.

Riedl, R. 1978. Order in Living Organisms: a Systems Analysis of Evolution. Wiley, Chichester.

Riley, R.M., W. Jin and G. Gibson. 2003. Contrasting selection pressures on components of the Ras-mediated signal transduction pathway in Drosophila. Mol Ecol 12: 1315–1323.

Rivera, M.C., R. Jain, J.E. Moore and J.A. Lake. 1998. Genomic evidence for two functionally distinct gene classes. Proc Natl Acad Sci U S A 95: 6239–6244.

Rivera, M.C. and J.A. Lake. 2004. The ring of life provides evidence for a genome fusion origin of eukaryotes. Nature 431: 152–155.

Rocha, E.P. 2006. The quest for the universals of protein evolution. Trends Genet 22: 412–416.

Rocha, E.P. and A. Danchin. 2004. An analysis of determinants of amino acids substitution rates in bacterial proteins. Mol Biol Evol 21: 108–116.

Rutherford, S.L. 2003. Between genotype and phenotype: protein chaperones and evolvability. Nat Rev Genet 4: 263–274.

Rutherford, S.L. and S. Lindquist. 1998. Hsp90 as a capacitor for morphological evolution. Nature 396: 336–342.

Sagan, L. 1967. On the origin of mitosing cells. J Theor Biol 14: 255–274.

Salathé, M., M. Ackermann and S. Bonhoeffer. 2006. The effect of multifunctionality on the rate of evolution in yeast. Mol Biol Evol 23: 721–722.

Scannell, D.R. and K.H. Wolfe. 2008. A burst of protein sequence evolution and a prolonged period of asymmetric evolution follow gene duplication in yeast. Genome Res 18: 137–147.

Sharkey, T.D., S. Yeh, A.E. Wiberley, T.G. Falbel, D. Gong and D.E. Fernandez. 2005. Evolution of the isoprene biosynthetic pathway in kudzu. Plant Physiol 137: 700–712.

Sharp, P.M. 1991. Determinants of DNA sequence divergence between Escherichia coli and Salmonella typhimurium: codon usage, map position, and concerted evolution. J Mol Evol 33: 23–33.

Sharp, P.M. and W.H. Li. 1987. The rate of synonymous substitution in enterobacterial genes is inversely related to codon usage bias. Mol Biol Evol 4: 222–230.

Slotte, T., T. Bataillon, T.T. Hansen, K. St Onge, S.I. Wright and M.H. Schierup. 2011. Genomic determinants of protein evolution and polymorphism in Arabidopsis. Genome Biol Evol.

Song, X., P. Jin, S. Qin, L. Chen and F. Ma. 2012. The evolution and origin of animal Toll-like receptor signaling pathway revealed by network-level molecular evolutionary analyses. PLoS One 7: e51657.

Stark, C., B.J. Breitkreutz, T. Reguly, L. Boucher, A. Breitkreutz and M. Tyers. 2006. BioGRID: a general repository for interaction datasets. Nucleic Acids Res 34: D535–539.

Subramanian, S. and S. Kumar. 2004. Gene expression intensity shapes evolutionary rates of the proteins encoded by the vertebrate genome. Genetics 168: 373–381.

Teichmann, S.A. 2002. The constraints protein-protein interactions place on sequence divergence. J Mol Biol 324: 399–407.

Thorne, J.L., N. Goldman and D.T. Jones. 1996. Combining protein evolution and secondary structure. Mol Biol Evol 13: 666–673.

Toft, C. and M.A. Fares. 2010. Structural calibration of the rates of amino acid evolution in a search for Darwin in drifting biological systems. Mol Biol Evol 27: 2375–2385.

Tokuriki, N. and D.S. Tawfik. 2009a. Chaperonin overexpression promotes genetic variation and enzyme evolution. Nature 459: 668–673.

Tokuriki, N. and D.S. Tawfik. 2009b. Stability effects of mutations and protein evolvability. Curr Opin Struct Biol 19: 596–604.

Toll-Riera, M., D. Bostick, M.M. Albà and J.B. Plotkin. 2012. Structure and age jointly influence rates of protein evolution. PLoS Comput Biol 8: e1002542.

Toll-Riera, M., J. Castresana and M.M. Albà. 2008. Accelerated evolution of genes of recent origin. pp. 45–59. *In*: P. Pontarotti (ed.). Evolutionary Biology From Concept to Application. Springer, New York.

Tuller, T., M. Kupiec and E. Ruppin. 2008. Evolutionary rate and gene expression across different brain regions. Genome Biol 9: R142.

Van de Peer, Y., J.S. Taylor, I. Braasch and A. Meyer. 2001. The ghost of selection past: rates of evolution and functional divergence of anciently duplicated genes. J Mol Evol 53: 436–446.

Vishnoi, A., S. Kryazhimskiy, G.A. Bazykin, S. Hannenhalli and J.B. Plotkin. 2010. Young proteins experience more variable selection pressures than old proteins. Genome Res 20: 1574–1581.

Vitkup, D., P. Kharchenko and A. Wagner. 2006. Influence of metabolic network structure and function on enzyme evolution. Genome Biol 7: R39.

von Mering, C., R. Krause, B. Snel, M. Cornell, S.G. Oliver, S. Fields and P. Bork. 2002. Comparative assessment of large-scale data sets of protein-protein interactions. Nature 417: 399–403.

Wagner, A. 2012. Metabolic networks and their evolution. Adv Exp Med Biol 751: 29–52.

Wagner, A. and D.A. Fell. 2001. The small world inside large metabolic networks. Proc Biol Sci 268: 1803–1810.

Wall, D.P., A.E. Hirsh, H.B. Fraser, J. Kumm, G. Giaever, M.B. Eisen and M.W. Feldman. 2005. Functional genomic analysis of the rates of protein evolution. Proc Natl Acad Sci U S A 102: 5483–5488.

Wang, G.Z. and M.J. Lercher. 2011. The effects of network neighbours on protein evolution. PLoS One 6: e18288.

Wang, M., Q. Wang, Z. Wang, X. Zhang and Y. Pan. 2013. The Molecular Evolutionary Patterns of the Insulin/FOXO Signaling Pathway. Evol Bioinform Online 9: 1–16.

Wang, Y., E.A. Franzosa, X.S. Zhang and Y. Xia. 2010. Protein evolution in yeast transcription factor subnetworks. Nucleic Acids Res 38: 5959–5969.

Wang, Z. and J. Zhang. 2009. Why is the correlation between gene importance and gene evolutionary rate so weak? PLoS Genet 5: e1000329.

Wilke, C.O. and D.A. Drummond. 2006. Population genetics of translational robustness. Genetics 173: 473–481.

Wilson, A.C., S.S. Carlson and T.J. White. 1977. Biochemical evolution. Annu Rev Biochem 46: 573–639.

Williams, T.A. and M.A. Fares. 2010. The effect of chaperonin buffering on protein evolution. Genome Biol Evol 2: 609–619.

Winter, E.E., L. Goodstadt and C.P. Ponting. 2004. Elevated rates of protein secretion, evolution, and disease among tissue-specific genes. Genome Res 14: 54–61.

Wolf, Y.I., L. Carmel and E.V. Koonin. 2006. Unifying measures of gene function and evolution. Proc Biol Sci 273: 1507–1515.

Wolf, Y.I., P.S. Novichkov, G.P. Karev, E.V. Koonin and D.J. Lipman. 2009. The universal distribution of evolutionary rates of genes and distinct characteristics of eukaryotic genes of different apparent ages. Proc Natl Acad Sci U S A 106: 7273–7280.

Wright, K.M. and M.D. Rausher. 2010. The evolution of control and distribution of adaptive mutations in a metabolic pathway. Genetics 184: 483–502.

Wright, S.I., C.B. Yau, M. Looseley and B.C. Meyers. 2004. Effects of gene expression on molecular evolution in Arabidopsis thaliana and Arabidopsis lyrata. Mol Biol Evol 21: 1719–1726.

Wu, X., X. Chi, P. Wang, D. Zheng, R. Ding and Y. Li. 2010. The evolutionary rate variation among genes of HOG-signaling pathway in yeast genomes. Biol Direct 5: 46.

Wyckoff, G.J., C.M. Malcom, E.J. Vallender and B.T. Lahn. 2005. A highly unexpected strong correlation between fixation probability of nonsynonymous mutations and mutation rate. Trends Genet 21: 381–385.

Xia, Y., E.A. Franzosa and M.B. Gerstein. 2009. Integrated assessment of genomic correlates of protein evolutionary rate. PLoS Comput Biol 5: e1000413.

Yang, D., Y. Jiang and F. He. 2009. An integrated view of the correlations between genomic and phenomic variables. J Genet Genomics 36: 645–651.

Yang, J., Z. Gu and W.H. Li. 2003. Rate of protein evolution versus fitness effect of gene deletion. Mol Biol Evol 20: 772–774.

Yang, J.R., B.Y. Liao, S.M. Zhuang and J. Zhang. 2012. Protein misinteraction avoidance causes highly expressed proteins to evolve slowly. Proc Natl Acad Sci U S A 109: E831–840.

Yang, L. and B.S. Gaut. 2011. Factors that contribute to variation in evolutionary rate among Arabidopsis genes. Mol Biol Evol 28: 2359–2369.

Yu, H.S., Y.H. Shen, G.X. Yuan, Y.G. Hu, H.E. Xu, Z.H. Xiang and Z. Zhang. 2011. Evidence of selection at melanin synthesis pathway loci during silkworm domestication. Mol Biol Evol 28: 1785–1799.

Zhang, J. and X. He. 2005. Significant impact of protein dispensability on the instantaneous rate of protein evolution. Mol Biol Evol 22: 1147–1155.

Zhang, L. and W.H. Li. 2004. Mammalian housekeeping genes evolve more slowly than tissue-specific genes. Mol Biol Evol 21: 236–239.

Zhang, P., Z. Gu and W.H. Li. 2003. Different evolutionary patterns between young duplicate genes in the human genome. Genome Biol 4: R56.

Zhang, Z. and J. Parsch. 2005. Positive correlation between evolutionary rate and recombination rate in Drosophila genes with male-biased expression. Mol Biol Evol 22: 1945–1947.

Zhou, T., D.A. Drummond and C.O. Wilke. 2008. Contact density affects protein evolutionary rate from bacteria to animals. J Mol Evol 66: 395–404.

Zuckerkandl, E. 1976. Evolutionary processes and evolutionary noise at the molecular level. I. Functional density in proteins. J Mol Evol 7: 167–183.

Zuckerkandl, E. and L. Pauling. 1965. Evolutionary divergence and convergence in proteins. Evolving Genes and Proteins 97: 166.

Chapter 8

The Network Framework of Molecular Evolution

Ludovica Montanucci, Hafid Laayouni and
Jaume Bertranpetit *

8.1 Introduction

The difficulty in the functional understanding of the genome
lies not only in the high number of functional elements but
also in the fact that they interact in entangled pathways,
giving rise to global properties whose relationships with
the specific role and function of each element are difficult
to discern. So, increasingly, the study of the genome
is approached with a perspective that, along with the
properties and functions of single genes, aspects of their
interactions are also taken into account. A straightforward
way to study an interacting system is to represent it through
a network. Hence network biology has received much
interest and interdisciplinary efforts, combining biological
problems with tools provided from physics or computer

IBE, Evolutionary Biology Institute (CSIC-Pompeu Fabra University)
  UPF-PRBB, Doctor Aiguader 88, 08003 Barcelona, Catalonia, Spain.
*Corresponding author

sciences (Barabási and Oltvai 2004; Laghaee et al. 2005; Barabási et al. 2011). Among the biomolecular ones, gene regulatory networks, protein interaction networks, and metabolic networks represent key systems that describe the basic mechanisms that govern the life and functioning of the cell.

To characterize the structure of these networks, tools provided by graph theory have been widely exploited. Graph theory provides a general theoretical framework that allows studying the structure and topology of large networks independently of the system that they model. It allows the computation of parameters that characterize the structure of the networks, to classify networks according to these parameters, and, when possible, to propose "generating rules", that drive the growth of the network and that can account for its observed structure (Dorogovtsev and Mendes 2003). Indeed the description of the topology itself should be able to reveal at least some of the principles and constraints of its evolution. The description of network topology can be carried out at a global scale through the determination of parameters that are computed taking into account the network as a whole. Parameters such as the degree (connectivity) distribution or the average shortest distance between nodes, allow characterizing and classifying the global organization of the network. Alternatively, one can focus on identifying local structures that are over-represented in the network. These structures, which are called network motifs and are composed of an arbitrary but usually small number of nodes, do not give information directly about the general structure of the network but can inform about its local organization and constituent parts (Alon 2007). The over-representation of these motifs puts forward the hypothesis that these may reflect a specific function and may be the topological units of evolution

(Wuchty et al. 2003). These topological descriptions have been drawn for many biomolecular networks and used to describe their global structure and allow delving into the evolutionary mechanisms that generated them.

A major fundamental feature of these molecular networks is that they are evolving. Their global structure varies through time as their components and connections are constantly added, changed and removed, and the components of many biological networks, such as genes and gene products, are themselves evolving entities. Given a biological pathway and its representation as a network, the question is how its structure (given by the functional interactions) affects the evolutionary history of its components by restraining their evolution (negative or purifying selection) or being at the base of innovations (positive or adaptive selection). Thus genes are considered as the encoding elements of a functional interactive process. The evolutionary patterns of the genes can be uncovered by the study of their inter- and intra-specific sequence variation that allows reconstructing their recent and remote evolutionary history and to reveal the selective pressures they underwent.

The importance of the studies that focus on the evolutionary properties of the network components themselves and relate them to the network structure is twofold. On the one hand, studying the selective pressures and evolutionary history of each gene in relation to its position and role within its interaction network represents a way of addressing an aspect of network evolution. On the other hand, molecular evolution can gain new insights by the introduction of the network information. Indeed while molecular evolution mainly explores variation at the gene level, biological functions, which are at the base of the phenotypic traits that are the units expected to be under natural selection, are usually the result of different interacting genes with uneven importance. These complex relationships make it difficult to predict the resulting phenotype from the analysis of single gene variation or, the other way around, to understand the molecular basis of an adaptive phenotypic change. A significant contribution in filling this gap could come from integrating the network-level of analysis in molecular evolution studies in order to gain insight into the genetic bases of adaptation.

Here we aim at reviewing the studies that have investigated the relationship between the evolutionary properties of protein-coding genes considered within the scope of interaction networks to seek how its topology and structure may drive the selection forces.

8.2 Network Representation of Biological Interactions

The first basic concept concerns the type of biological interactions. Biological entities, such as genes and proteins (gene products), simultaneously interact in many ways and the different types of interactions contribute at the same time to the global functioning of the system; however each type operates by different mechanisms and respond to different constraints. Therefore it may be worth treating them separately. Indeed, even if, ultimately, all the components in the cell are integrated in a single interacting system, different levels of description can be detached to make the problem of studying the evolution of genes within the entangled web of all the cellular interaction treatable and to focus on a particular aspect of the interactions to investigate its own properties. Moreover, the present knowledge of molecular biology is far from the whole accurate picture that would allow a global integration. Three types are usually considered through the study of three basic interaction types: the protein-protein interactions that capture all the physical contacts among the gene products (proteins) in the cell, the metabolic interactions that capture the concerted action of different gene products (enzymes) to produce essential components of the cell through biochemical reactions, and the gene regulatory interactions that capture the relationships between transcription factors and their target (regulated) genes. Arguably in the future the growing types of functional RNAs will have to be included.

A second and major issue concerns how to properly encode the chosen information about the interaction into a network. For many biologically relevant networks, different encodings (different definitions of what nodes and edges represent and the consequent network construction) are possible. Indeed nodes can represent genes, domains, or complexes; while edges may or may not have a defined orientation or directionality and may represent different types of relationships, even in the same network. A clear example is given by the metabolic network with which at least two alternative encodings are equally popular: one having metabolites as nodes and reactions as edges and the other having enzymes as nodes and the sharing of a common metabolite as edges. In fact, the type of encoding influences the questions that can be posed and the results that are consequently drawn (Montañez et al. 2010).

8.3 Distribution of Selection Over Biomolecular Networks

In the study of the evolution of genes within the context of their genomic interactions by the analysis of how different selective pressures distribute over biomolecular networks, two main routes have been pursued, differentiated by the size of the analyzed networks. A first set of studies has focused on a small set of genes interacting among themselves to perform a given function (herein referred to as "small-scale" studies). These studies can take advantage of the knowledge of the specific biological role of each gene within the network, favoring an interpretation in the light of the biological function served. Nonetheless, these studies are likely to reveal pathway specific patterns that may not be generalized to other pathways and, if the number of genes involved in the specific pathway is too small, they may lack statistical power to detect correlations with topological parameters. A second group of studies, made possible by the compilation of genomic maps of interactions, have investigated the evolutionary rates on a genome wide scale, allowing unprecedented global perspectives. The results of these studies, however, are hardly interpretable in the light of specific biological functions that emerge from the entangled web of interactions.

8.4 Characterizing the Evolutionary Properties of Genes

The history of the evolution of each individual gene, including the action of the selective pressures that acted on it, can be inferred from the analysis of the genetic variation among its sequences either from different species for its distant history or from different individuals of the same species for a more recent history.

It is mainly the inter-specific variations distribution over functional molecular networks that has been analyzed.

The study of inter-specific variation allows one to estimate the rates of sequence evolution and to detect deviations from neutrality, indicative of the past action of selective constraints (purifying selection, decelerated evolution) or positive selection (adaptive selection, accelerated evolution). For protein-coding DNA sequences, several parameters can be estimated and tests of deviation from neutrality have been implemented. These are available in computational packages (Pond et al. 2005; Yang 2007; Pybus et al. 2013) and the synonymous (d_S) and non-synonymous (d_N) substitution rates have been widely used. The first can be taken as the rate of neutral evolution of the genomic region, and the latter as the rate of functional evolution (given that these changes affect protein sequence). Their ratio (d_N/d_S, also called ω) allows detecting whether the non-synonymous changes have evolved at an accelerated or decelerated rate with respect to the neutral ones and hence infers the direction and the strength of the average selective constraint acting on the gene. It allows the safe comparison of the strength of selection on genes that lie in different genome regions, that may be evolving at very different rates due to non-selective causes such as local nucleotide composition (GC-content) or chromosomal location (telomeric versus centromeric).

Given a multiple-sequence alignment of orthologous genes from different species, a single d_N/d_S can be estimated for the entire gene sequence. This computation assumes the same evolutionary rate for all the codon sites along the sequence length and for all the branches in the phylogenetic tree of the gene in the considered species and it constitutes a widely used metric of selection (Eyre-Walker 2006; Pál et al. 2006). However, multiple d_N/d_S ratios can also be computed for different branches of the phylogenetic tree or for different classes of sites along the sequence.

The power of a unique d_N/d_S for a gene has been debated since its introduction. It was first interpreted as a proxy for protein dispensability (Wilson et al. 1977), but this interpretation has since been challenged (Pál et al. 2003). The computation of this parameter presupposes that the rate of evolution is constant along the sequence and over the entire divergence time, and this is a quite unrealistic assumption. It is known that the molecular structure of amino acids (i.e., *side-chain composition*) or their positions along the protein's primary structure, both affect the substitution rate, thus individual codons are expected to evolve at different rates. It is also unlikely that selective pressures have not varied during a time period as long as that required for the divergence of many species. Therefore more realistic multi-rate models have been developed to account for rate variation along the sequence (Yang et al. 2000) and along branches (Yang 1997; Zhang et al. 2005), depending on the type of amino acid (Delport et al. 2010). Even if these models are more realistic, they depend on a greater number of parameters and, to achieve adequate accuracy, an appropriate amount of sequence data (in terms of species represented and number of substitutions, that is, amount of divergence in the considered species) is required as the complexity of the model increases.

Unlike the simplistic, single-rate model, these more realistic models may supply tests to detect events of positive selection. However, in the absence of signatures of positive selection, it is still valuable to have a single estimate that summarizes the direction and the average strength of the selection on the protein sequence during the divergence of the considered species. This single d_N/d_S is an informative measure of the evolutionary constraint on the protein sequence.

When investigating the relationship of the evolutionary rates with network properties, all the factors possibly influencing evolutionary rates have to be considered in order to properly isolate the independent relationships between evolutionary rates and network structure. Expression level has long been known as a major factor influencing evolutionary rates (Rocha and Danchin 2003): highly expressed proteins tend to evolve slowly. The most accepted explanation is the translational robustness hypothesis, according to which pressure for a low rate of translational errors will constrain sequence evolution. Expression level has been proposed to be the unique determinant driving all the observed variation (Drummond et al. 2006). Nonetheless, independent contributions of other genomic determinants of evolutionary rates have been found including protein length, protein age, protein structure, essentiality and protein dispensability (Wall et al. 2005; Bloom et al. 2006). The relationships among all these factors are entangled because many of them are correlated among themselves (Krylov et al. 2003).

At the intra-specific level, comparison from DNA sequences and SNP data mainly focuses on the detection of positive selection. Several methods use the comparison within and between species, such as the HKA test. Other classical methods use population differentiation index (Fst) or summary statistics of allele frequency spectra. Many new methods to detect selection rely on the use of linkage disequilibrium and haplotype structure (Nielsen 2005). The time scale is an important issue when considering a method for detecting selection, ranging from millions of years in the case of divergence (d_N/d_S) to a few thousands of years in the case of linkage disequilibrium decay methods such as those based on the long range haplotype (iHS and others). A large number of tests have been implemented for studying intraspecific variation in humans from data of the 1000 genome project (Pybus et al. 2013).

8.5 Think Locally: Small-Scale Functional Networks

Early attempts to couple the analysis of the molecular evolution of genes with the knowledge of their network interactions explored small and well known pathways. In these cases, interactions were determined on the basis of well established molecular knowledge on the process.

8.5.1 Inter-specific Analysis

The first study was the anthocyanin biosynthetic pathway, a metabolic pathway with only 6 genes from 3 plant species (Rausher et al. 1999). Lately, other metabolic biosynthetic pathways (Flowers et al. 2007; Livingstone and Anderson 2009; Yang et al. 2009; Ramsay et al. 2009; Montanucci et al. 2011) and a signaling pathway (Riley et al. 2003; Alvarez-Ponce et al. 2009) have been analyzed. These pathways share the property of being composed of a small number of genes functionally related, and the role of each gene within the system is well known, allowing the construction of a network structure from primary biochemical research articles. These works couple the study of the molecular evolution of the genes of a pathway with a network-level analysis to detect whether differences in the selective pressures acting on genes is due, in part, to constraints imposed by the pathway structure. These comparative methods mostly estimate the strength of purifying selection and are rarely able to detect adaptive evolution.

For many of the analyzed pathways, it has been found that upstream genes in the pathway tend to evolve slower than those downstream due to a stronger selective constraint (Rausher et al. 1999; Lu and Rausher 2003; Riley et al. 2003; Rausher et al. 2008; Livingstone and Anderson 2009; Ramsay et al. 2009). A proposed explanation for this observed pattern is that upstream genes are under stronger purifying selection because they are more pleiotropic than those downstream, affecting a greater number of end products, because they are likely to be above branch points and hence involved in the synthesis of more products than those downstream (Rausher et al. 1999).

However, this gradient of decreasing purifying selection along a pathway has not always been found. No significant evidence for it was found in the case of gibberilin metabolic biosynthetic pathway (Yang et al. 2009), while a gradient in the opposite direction (downstream genes are under stronger purifying selection) was found for the signaling pathway of the orthologous insulin/TOR pathway structure both in Drosophila (Alvarez-Ponce et al. 2009) and in vertebrates (Alvarez-Ponce et al. 2011) and in a metabolic biosynthetic pathway, the N-glycosylation pathway (Montanucci et al. 2011).

Another measure that stresses the position of a genes in relation to pathway branch points is the "Pathway Pleiotropy Index" (PPI) (Ramsay et al. 2009), which counts groups of enzymes between pathway branch points: it is an ordinal number and groups of enzymes between two consecutive branch points are given the same position and hence the same number. In a dataset composed of 40 genes, PPI correlates with evolutionary rates better than simple pathway position and this supports the hypothesis that pleiotropy has a significant effect on evolutionary rates. The importance of branch points in metabolic pathways further emerges from two other studies in which adaptive evolution is found to happen on genes located at branch points (Flowers et al. 2007) and enzymes at branch points are found to be under stronger purifying selection (Yang et al. 2009). This supports the notion that branch points in metabolic pathways are critical to evolution, probably because they have great impact on flux (Laportes et al. 1984; Olson-Manning et al. 2013).

8.5.2 Intra-specific Analysis

Even if most studies used inter-specific variation data to study molecular evolution for longer time-scale periods, some recent work has investigated the impact of network structure on recent positive selection using intra-specific data (Luisi et al. 2012; Dall'Olio et al. 2012). In the first study, authors used human population genetics data to study the molecular evolution of genes involved in the human insulin/TOR signal transduction pathway and used single nucleotide polymorphism to assess variation in human populations. To estimate positive selection, they used methods based on linkage disequilibrium and differences in allele frequencies between populations. The main result of this work is related to the incidence of positive selection and the position of the genes (central or peripheral) in the network: genes evolving under positive selection encode proteins that have a significantly higher connectivity within the insulin/TOR network. These results were significant when protein-protein interactions were considered, and also when metabolic and transcriptional activation interactions were taken into account. Thus, the impact of positive selection on the insulin/TOR pathway genes is related to the centralities of their encoded products in the interaction network, with genes acting at the center of the network being more likely to evolve under positive selection.

In the second study, Dall'Olio et al. (2012) focused on the metabolic pathway of N-glycosylation. Results showed that genes in the downstream part of the pathway are more likely to show a signature of population differentiation, and that events of positive selection are frequent on genes that are known to be at bifurcation points, and that are identified as being in a key position by a network-level analysis.

These studies point to an interesting feature: positive selection at the intraspecific level seem to concentrate in fundamental parts of the network, in central or bifurcating places, with higher connectivity, and thus with high likelihood of having strong pleiotropy and phenotypic effect.

8.5.3 Overview on small-scale studies

Datasets for small-scale studies are built on biological knowledge of the process and comprise a limited number of genes and are usually characterized by a high quality of genetic sequences and a good confidence in the biochemical interactions used to construct the networks (Dall'Olio et al. 2011).

All these studies suggest that a small part of the variation of evolutionary rates can be accounted for by the structure of their functional network; however different patterns have been found for different pathways and different species sets. The emerging results point to a structure of the strength of purifying selection, decreasing in the functional sense of the pathway and thus being stronger for key functional positions, as expected; for positive selection, results at the

191

interspecific level are not clear without a common trend in the studied pathways, while at the intraspecific level, surprisingly, it seems to concentrate in key positions of the pathway (and thus in hubs of the network).

8.6 Think Globally: Whole-genome Networks

In addition to small-scale pathways, interest in whole-genome networks grew as soon as the first compilations of high amounts of data were made available. Many types of biological networks can be defined and studied in an evolutionary context. Here we will focus on three main network types, the protein-protein interaction, the metabolic and the transcription factor networks. The general approach adopted in such large-scale studies has been first, to study the topology of the network in order to describe its structure and characterize it and second, to explore the relationships between topological characteristics of the network and footprints of selection (purifying or adaptive) on genes in order to uncover patterns of evolution.

8.6.1 Protein-protein interaction network

Whole-genome protein-protein interaction (PPI) networks have received great interest; nodes in the protein interaction networks represent proteins and edges represent physical interactions among them (Nooren and Thornton 2003). Since the development of the yeast two-hybrid technique, high throughput determinations of physical interactions among proteins allowed proteome-scale reconstructions of the interactome for many species.

When the topology of these networks was investigated, a global organization compatible with a scale-free model was found (see Barabási and Oltvai 2004; Yamada and Bork 2009). The scale-free topology is defined by the shape of the connectivity which follows a power law. Therefore, the discovery of this global organization of the protein interaction network brings to attention the role and importance of node connectivity. A key question is whether the topology itself confers some properties to the system and what may be its possible functional implications in case of the PPI networks. The fact that the scale-free organization emerges independently in many biomolecular (and also non-biological) networks, also suggests that this topology arises from a self-organizing process, which could be the influence or even be the result of a selective process.

An implication of the scale-free connectivity distribution is the existence of a relatively small number of nodes within the network that have a high number of connections (named "hubs") that play a crucial role in maintaining the overall network structure. Several studies have then investigated their biological features to discover possible functions and characteristics of these highly connected proteins. Indeed it has long been debated whether hub proteins have any special features, for example as being more essential to the cell, where an essential protein is defined as one whose deletion is lethal. Here we will review its relationship to selective forces.

8.6.1.1 Connectivity and Indispensability

Among many other biological properties, a special interest has been given to the study of protein indispensability, experimentally measured as the viability following gene deletion. When it was investigated along with connectivity in the yeast protein-interaction network, it was found that highly connected proteins are three times more likely to be essential than less-connected ones (Jeong et al. 2001; Wuchty 2004). These findings, known as the centrality-lethality rule (He and Zhang 2006), suggest that indispensability of proteins is due, at least in part, to their position within the network or, in other words, it has a topological component. When hub proteins are removed (for example due to gene deletion) the network is quickly disrupted, while, on the contrary, the removal of a low-connected node is generally tolerated, in the sense that it hardly affects the global network structure. This topological property is likely to contribute to the robustness of the protein interaction network with a wide variety of importance in biological functions.

8.6.1.2 Connectivity and Evolutionary Rates

If highly connected proteins are more essential and their disruption is more likely to be lethal, they should be under stronger negative selection and this should be identifiable in the rates of their sequence evolution. The relationship between evolutionary rates and connectivity (see Box 1) in the protein interaction network has then been largely investigated to test the hypothesis of mutational robustness, and, more generally, to address the general question of how selection differently targets the components of an interacting system.

Fraser et al. (2002) found a negative correlation between connectivity and the protein evolutionary rate, estimated as the number of substitutions per amino acid site between orthologous protein sequences of *S. cerevisiae*. This means that proteins that interact with many other proteins tend to evolve more slowly than less connected ones. Two different

Graphs

The mathematical structure widely adopted to model interacting systems is the graph, which represents the analytical formalization of the general and more intuitive concept of network. Graphs are made of *nodes* which represent entities (in case of biological networks they can represent proteins, genes, enzymes, compounds...) and *edges* (also named links or connections) which represent the interactions among the entities (such as regulation or metabolic relationships, physical protein interaction).

Centrality measures

To characterize the position of a node with respect of the whole network, different measures, named centrality measures, have been defined, each one capturing different aspects of the position of a node. Here are presented those that have been mainly used within the context of the characterization of biomolecular networks.

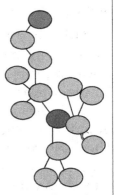

- **Degree centrality** of a node is defined as the number of its connections (also named connectivity of a node). The green node has a connectivity of 1 while the orange and the red nodes have a connectivity of 4 and 3 respectively. Connectivity can be normalized by dividing it by the total number of connections in the whole network.

- **Betweenness centrality** of a node is defined as the fraction of all shortest paths between all pairs of nodes in the network that pass through that node. It captures whether the nodes occupy "bridge" positions. The orange and red nodes are both in bridge positions for communicating the two parts of the network and thus have the highest values of betweenness.

Box 1 Figure*

- **Closeness centrality** of a node is defined as the reciprocal of its average distance with all the nodes in the network. Intuitively it tells whether the node lies in the center or in the periphery of the network. The green node lye at the periphery of the network and has low value of closeness centrality.

Scale-free networks

Scale-free networks (first described in Barabási and Albert 1999) are defined by the shape of their connectivity distribution which is the distribution of the connectivities of all the nodes of the network. The connectivity distribution of a scale-free graph, which differs from that of random ones, is best approximated by a power law function. This implies that the majority of the nodes have few connections and there are few nodes with many connections (highly connected nodes). These highly connected nodes are more than expected in random graphs and are called hubs.

Color image of this figure appears in the color plate section at the end of the book.

hypotheses are considered to account for this observed correlation. The first is that highly connected protein may be more important for the cell by exhibiting a greater effect on fitness, in agreement with the mutational robustness hypothesis introduced by Jeong et al. (2001). Alternatively, highly connected proteins may have a larger proportion of their structure involved in their function and thus may be under an overall higher constraint. A partial correlation analysis led the authors to conclude that the correlation between connectivity and evolutionary rates is not mediated by the fitness, but it is due to a higher proportion of the protein structure involved in its function.

This correlation between evolutionary rates and connectivity has long been debated. Some independent research provided further evidence for it (Wuchty 2004), however it has not been confirmed (or only partially) by other studies in which different interaction data sets and different evolutionary distances have been considered. These discrepancies point out the importance of a critical consideration of the results in light of the available interaction data, because there are concerns about the networks' reconstruction associated with the experimental techniques used (Zhu et al. 2007). In fact, such a bias in some experimental methods for the determination of protein-protein interactions has been claimed to be the basis of the conflicting results found in Fraser et al. (2002) and Jordan et al. (2003) for the correlation between evolutionary rates and connectivity (Bloom and Adami 2003). In this last work the authors sought the correlation between connectivity and evolutionary rates in different interaction data sets and discovered that this correlation is stronger when including datasets erring towards counting more interactions for abundant proteins. Indeed, since protein expression level is known to be a major factor constraining the substitution rate of protein sequences in bacteria (Drummond et al. 2006), with highly expressed proteins showing lower rates of evolution, the bias in interaction datasets could drive the correlation between connectivity and evolutionary rates. This last observation points out the importance of considering the effects of other genomic determinants which are known to influence evolutionary rates when trying to relate them to networks properties. However, even when the effect of expression levels is ruled out, the adoption of different compilations of the yeast protein interaction network results in inconsistent correlation patterns (Fraser and Hirsh 2004).

In summary, while the role of connectivity in constraining protein evolutionary rates in protein interaction networks may depend on the datasets used in the analysis and thus remains controversial, most authors accept a correlation between connectivity and evolutionary rates.

Beyond connectivity, other measures that describe the position of a gene within the network in which it participates have been used (see Box 1). Along with connectivity, two centrality measures, betweenness and closeness centrality, have been explored in three different eukaryotic networks and it appeared that betweenness centrality correlates with evolutionary rate estimated from the DNA sequence as d_N and d_N/d_S (Hahn and Kern 2005) independently of the connectivity. So, independently of the number of connections of each protein in the interaction network, those that are central according to the betweenness measure, evolve more slowly. Proteins with high betweenness (behaving as information bridges) are more likely to drive the cross-talk between different parts of the network and to have higher pleiotropic effects. The fact that these proteins are found to evolve under stronger constraint is in agreement with Fisher's pleiotropy hypothesis and has also been invoked to explain patterns of the distribution of evolutionary constraint in small scale networks.

8.6.1.3 Positive Selection in Protein-Interaction Networks

Besides analyzing different rates of sequence evolution, the position in the network of genes having evolved under positive selection has also been investigated. The fraction of genes that show this signature is generally relatively small but they are interesting because they are at the basis of new adaptations and innovations. Kim and coworkers sought where, within the human protein-protein interaction network, events of positive selection have been allowed to take place within the human protein-protein interaction network (Kim et al. 2007) during the human and chimpanzee divergence. The probability of a gene having undergone an event of positive selection was approximated through the likelihood ratio derived from the d_N/d_S test. It was found that events of positive selection tend to have occurred at the periphery of the network, which is claimed to correspond to the physical periphery of the cell. Adaptation at the periphery of the network at the interspecific level would mean that adaptations tend to occur in low-connected nodes and thus in genes coding for proteins of low importance, lowly pleiotropic and less essential.

An alternative way of considering the position of a protein within the interaction network is to consider its participation and position in relation to modules, which can be defined as groups of highly connected genes that perform a given function. The rate of evolution has been estimated by d_N/d_S and it emerged that hub proteins that connect different modules have been more free to evolve than those whose connections are principally within modules (Fraser 2005). This finding hints toward the idea of an evolution through tinkering in which modules are quite constrained and conserved while their co-option is more prone to be under selection.

8.6.2 Metabolic interaction networks

Knowledge on metabolic reactions in model organisms has been drawn from decades of biochemical research and its integration with the currently available whole-genome information is allowing the reconstruction of a single organism-scale metabolic network (Förster et al. 2003; Duarte et al. 2004; Duarte et al. 2007) encompassing the metabolic machinery at a system level. As the reconstruction of metabolic networks encompasses a huge amount of knowledge on the biochemistry of the involved processes, metabolic networks happen to be more tightly identifiable with the performed function and more easily interpreted in light of the function of each enzyme-coding gene and its role within the global process.

The information comprised in metabolic networks is manifold: information about reactions, about proteins (enzymes catalyzing the reactions or their encoding genes) and about metabolites (small molecules and compounds transformed during the metabolic reactions). Given this coupled information, there are many alternative ways of encoding metabolic information through graph structures and each representation gives a different focus to the analysis, capturing different aspects of the interaction organization of the metabolism machinery. Two main representations have been widely adopted (Wagner and Fell 2001): the first one is the substrate graph, in which nodes represent substrates and edges indicate their co-occurrence in the same reaction, usually as a substrate (node) linked by an enzyme (edge) to the product (another node); the second one is the reaction graph, in which nodes represent reactions and edges indicate shared compounds. Modified versions of these two representation classes can be built depending on how reversible reactions are treated; thus irreversible reactions can be represented either by directed (Ma and Zeng 2003a) or non-directed (Wagner and Fell 2001) edges while reversible reactions can be represented either by non-directed edges (Ma and Zeng 2003a; Wagner and Fell 2001) or by a pair of directed edges with opposite directions (Jeong et al. 2000; Light and Kraulis 2004) and also depending on whether the most highly connected metabolites (also named currency metabolites) are included (Jeong et al. 2000) or excluded (Wagner and Fell 2001; Ma and Zeng 2003a; Light and Kraulis 2004) from the graph.

Much research has been devoted to the study of the global organization of metabolic networks with a special interest in whether global topological features were selected during evolution and whether they provided the system with some specific biological function. When the topology of the metabolic networks, represented as a substrate graph, was studied, three main characteristics of its organization were described: its small-world property (Wagner and Fell 2001; Ma and Zeng 2003b), its scale-free organization (Jeong et al. 2000; Ma and Zeng 2003b) and its hierarchical organization (Ravasz et al. 2002). The fact that it has characteristics compatible with a small-world model coarsely implies that most pairs of nodes can be connected through a relatively short path. It has been speculated that these short paths between nodes may keep a high pace of information spread within the network and this may result in an advantage by enabling the quick reestablishment of the required concentration of metabolites in case of perturbations of the system.

A representation of the metabolic network that consists of genes (enzyme-coding genes) as nodes can be highly informative to study how selective pressures on the genes encoding for the enzymes are distributed throughout the network. Edges are established between nodes if the two corresponding enzymes share a common metabolite either as substrate or product. With this representation it is possible to consider the relative position of each enzyme-coding gene within the network and relate it to its evolutionary properties.

8.6.2.1 Evolutionary Rates and Centrality Measures

Hahn et al. (2004) found no relationship between connectivity and the rate of protein evolution, measured as amino acid divergence (d_N), in the *E. coli* metabolic network. However, a negative correlation has been reported when a bigger set of enzymes were analyzed in the *Drosophila* and yeast metabolic networks, with the evolutionary rates measured as d_N/d_S. In these networks, highly connected genes have been shown to evolve at slower rates of evolution indicating a greater selective constraint acting on them (Vitkup et al. 2006; Greenberg et al. 2008).

The connectivity measure for each node of the network primarily captures information about its neighborhood, through the number of nodes that are directly connected to it, without providing direct knowledge about how the node is placed in relation to the rest of the network. When other measures of centrality have been investigated, betweenness centrality has been found to better correlate to the evolutionary rates than a simple measure of degree (Lu et al. 2007).

To our knowledge, there is no study trying to relate the metabolic network with either purifying or adaptive selection at the intraspecific level.

8.6.3 Transcription Factor Network

The regulatory relationships among transcription factors (TFs) and their targets, which in turn can also encode transcription factors, can be represented through a network called transcription factor (or gene regulatory) network. A natural representation of the TF network, which retains the unequal relationship among genes (either regulating or being regulated), is through a directed graph. Nodes represent genes, either TFs or their targets, and edges are directed arrows (arcs) stemming from the transcription factor-encoding (regulating) gene and pointing to the target (regulated) gene, which, in turn, may encode a transcription factor. For directed graphs, two different connectivities can be computed: the in-degree, which is the number of incoming edges upon a node, and the out-degree, which is the number of outgoing edges from a node; these two connectivity measures have a straightforward interpretation.

A major problem in studying TF network stems from the fact that available network reconstructions may suffer from incompleteness since TF-target associations are highly context dependent (Papp and Oliver 2005) and experimental methods just test a few of them: arguably the information available now is largely incomplete. Nevertheless, many large scale compilations of TF maps have been produced and the global topology of the yeast TF network has been analyzed to understand features of its architecture and principles of its evolution (Babu et al. 2004; Rodriguez-Caso et al. 2005).

TFs are organized into a pyramid-shape hierarchical structure (Ma et al. 2004; Yu and Gerstein 2006). Few TFs, the master regulators, lie at the top of the pyramid and are those that are not regulated by other TFs and that receive information through protein-protein interaction.

8.6.3.1 Evolutionary rates and Centralities

The analysis of the yeast TF network showed that TF hubs do not evolve at lower rates than non hubs and hence are not under stronger selective constraint (Evangelisti and Wagner 2004). Neither are TF hubs more conserved across genomes when their presence/absence was investigated in the TF networks from 175 prokaryotic genomes (Babu et al. 2006). Another study on a bigger dataset, however, found a significant correlation between the rate of protein evolution and centrality, such that more central transcription factors tend to evolve faster (Jovelin and Philips 2009). They also found that centrality was also positively correlated with expression variability, suggesting that the higher rate of divergence among central transcription factors may be due to their role in controlling information flow and may be the result of adaptation to changing environmental conditions (Jovelin and Philips 2009).

No general agreement has therefore been achieved for the role of topological parameters in constraining the TF genes and further studies are needed. While there is a general support to the view that hub proteins in the protein-protein interaction network tend to evolve more slowly than non-hubs, in TF networks it is not that clear.

8.6.3.2 Gene co-expression networks

Another perspective of gene regulation is through the analysis of co-expression networks, where the connection between two genes is given by their co-expression in the same tissues and conditions. Jordan et al. (2004) found that the human gene co-expression network, derived from tissue-specific expression profiles, shows scale-free properties that imply evolutionary self-organization via preferential node attachment, where genes with numerous co-expressed partners (the hubs of the co-expression network) evolve more slowly on average than genes with fewer co-expressed partners, and genes that are co-expressed show similar rates of evolution.

8.7 Concluding Remarks

A key issue in molecular evolution is how selection, which acts on phenotypes, shapes the evolution of the genes that contribute to the phenotype through their collective and concerted action. The results of the studies reviewed here stress the usefulness of encoding molecular systems through biomolecular networks to capture the collective action of the genes that gives rise to the phenotype as an emergent property. In particular they demonstrate that the position of a gene within its biomolecular network accounts for a part of the variability in evolutionary rates between genes. In other words, this demonstrates that network organization imposes constraints on the evolution of its constituent genes and that this holds both for purifying and adaptive selection.

However universal patterns and general principles have been hardly derived despite several independent pieces of evidence of the constraint imposed by network structure on genes' evolution, which have been found for different network types and at different scales. A possible cause for this puzzling scenario stems from the fact that the constraints imposed by network structure seem to depend on the specific types of network considered, its size and, in case of small-scale networks, on the specific system. A certain degree of discrepancy between the obtained results could also reflect the poor quality of the network determinations, in particular for whole-genome approaches. Indeed a major limitation for these studies has been the poor confidence on present network structures, in terms of false positives and low coverage and the lack of species-specific networks. Advances in this area will certainly help in clarifying the impact of network structure in shaping the evolution of genes. A determinant factor seems to be the time depth of the compared genomes: while interspecific comparisons seem to point to an accumulation of both purifying selection and adaptive selection in the hubs in some of the analysis, intraspecific comparisons seem to reflect an increase in adaptive selection in highly connected genes. This point will have to be further analyzed in the future.

In fact, much knowledge has been and will be derived from considering the topological organization of the networks. In the future, a deeper understanding will be provided by taking into account not only the topology of the network, but also their dynamics and their functioning in space and time. However, much experimental work is still needed to gather all the information to allow such kind of analysis. At the very end, these studies are essential for the understanding the genetic bases of complex adaptation, in which the network approach will contribute significantly.

References

Alon, U. 2007. Network motifs: theory and experimental approaches. Nat Rev Genet 8: 450–461.

Alvarez-Ponce, D., M. Aguadé and J. Rozas. 2011. Comparative genomics of the vertebrate insulin/TOR signal transduction pathway: a network-level analysis of selective pressures. Genome Biol Evol 3: 87–101.

Alvarez-ponce, D., M. Aguadé and J. Rozas. 2009. Network-level molecular evolutionary analysis of the insulin/TOR signal transduction pathway across 12 Drosophila genomes. Genome Res 19: 234–242.

Babu, M.M., N.M. Luscombe, L. Aravind, M. Gerstein and S.A. Teichmann. 2004. Structure and evolution of transcriptional regulatory networks. Curr Opin Struct Biol 14: 283–291.

Barabási, A.-L. and Z.N. Oltvai. 2004. Network biology: understanding the cell's functional organization. Nat Rev Genet 5: 101–113.

Barabási, A.L., N. Gulbahce and J. Loscalzo. 2011. Network Medicine: A Network-based Approach to Human Disease. Nat Rev Genet 2: 56–68.

Bloom, J.D. and C. Adami. 2003. Apparent dependence of protein evolutionary rate on number of interactions is linked to biases in protein-protein interactions data sets. BMC Evol Biol 3: 21.

Bloom, J.D. D.A. Drummond, F.H. Arnold and C.O. Wilke. 2006. Structural Determinants of the Rate of Protein Evolution in Yeast. Mol Biol Evol 23: 1751–1761.

Dall'Olio, G.M., H. Laayouni, P. Luisi, M. Sikora, L. Montanucci and J. Bertranpetit. 2012. Distribution of events of positive selection and population differentiation in a metabolic pathway: the case of asparagine N-glycosylation. BMC Evol Biol 12: 98.

Dall'Olio, G.M., B. Jassal, L. Montanucci, P. Gagneux, J. Bertranpetit and H. Laayouni. 2011. The annotation of the asparagine N-linked glycosylation pathway in the Reactome database. Glycobiology 21: 1395–1400.

Delport, W., K. Scheffler, G. Botha, M.B. Gravenor, S.V. Muse, L. Sergei and S.L.K. Pond. 2010. CodonTest: Modeling Amino Acid Substitution Preferences in Coding Sequences. PLoS Comput Biol 6: e1000885.

Dorogovtsev, S.N. and J.F.F. Mendes. 2003. Evolution of Networks. Oxford University Press.

Drummond, D.A., A. Raval and C.O. Wilke. 2006. A single determinant dominates the rate of yeast protein evolution. Mol Biol Evol 23: 327–337.

Duarte, N.C., S.A. Becker, N. Jamshidi, I. Thiele, M.L. Mo, T.D. Vo, R. Srivas and B.Ø. Palsson. 2007. Global reconstruction of the human metabolic network based on genomic and bibliomic data. Proc Natl Acad Sci U S A 104: 1777–1782.

Duarte, N.C., M.J. Herrgård and B.Ø. Palsson. 2004. Reconstruction and validation of Saccharomyces cerevisiae iND750, a fully compartmentalized genome-scale metabolic model. Genome Res 14: 1298–1309.

206

Evangelisti, A.M. and A. Wagner. 2004. Molecular evolution in the yeast transcriptional regulation network. J Exp Zool B Mol Dev Evol 302: 392–411.

Eyre-Walker, A. 2006. The genomic rate of adaptive evolution. Trends Ecol Evol 21: 569–575.

Flowers, J.M., E. Sezgin, S. Kumagai, D.D. Duvernell, L.M. Matzkin, P.S.Schmidt and W.F. Eanes. 2007. Adaptive evolution of metabolic pathways in Drosophila. Mol Biol Evol 24: 1347–1354.

Fraser, H.B., A.E. Hirsh, L.M. Steinmetz, C. Scharfe and M.W. Feldman. 2002. Evolutionary rate in the protein interaction network. Science 296: 750–752.

Fraser, H.B. and A.E. Hirsh. 2004. Evolutionary rate depends on number of protein-protein interactions independently of gene expression level. BMC Evol Biol 4: 13.

Fraser, H.B. 2005. Modularity and evolutionary constraint on proteins. Nat Genet 37: 351–352.

Förster, J., I. Famili, P. Fu, B.Ø. Palsson and J. Nielsen. 2003. Genome-scale reconstruction of the Saccharomyces cerevisiae metabolic network. Genome Res 13: 244–253.

Greenberg, A.J., S.R. Stockwell and A.G. Clark. 2008. Evolutionary constraint and adaptation in the metabolic network of Drosophila. Mol Biol Evol 25: 2537–2546.

Hahn, M.W., G.C. Conant and A. Wagner. 2004. Molecular Evolution in Large Genetic Networks: Does Connectivity Equal Constraint? J Mol Evol 58: 203–211.

Hahn, M.W. and A.D. Kern. 2005. Comparative genomics of centrality and essentiality in three eukaryotic protein-interaction networks. Mol Biol Evol 22: 803–806.

He, X. and J. Zhang. 2006. Why Do Hubs Tend to Be Essential in Protein Networks? PLoS Genet 2: e88.

Jeong, H., S.P. Mason, A.L. Barabási and Z.N. Oltvai. 2001. Lethality and centrality in protein networks. Nature 411: 41–42.

Jeong, H.B. Tombor, R. Albert, Z.N. Oltvai and A.L. Barabási. 2000. The large-scale organization of metabolic networks. Nature 407: 651–654.

Jordan, I.K., L. Mariño-Ramírez, Y.I. Wolf and E.V. Koonin. 2004. Conservation and Coevolution in the Scale-Free Human Gene Coexpression Network. Mol Biol Evol 21: 2058–2070.

Jordan, I.K., Y.I. Wolf and E.V. Koonin. 2003. No simple dependence between protein evolution rate and the number of protein-protein interactions: only the most prolific interactors tend to evolve slowly. BMC Evol Biol 3: 1.

Jovelin, R. and P.C. Phillips. 2009. Evolutionary rates and centrality in the yeast gene regulatory network. Genome Biol 10: R35.

Kim, P.M., J.O. Korbel and M.B. Gerstein. 2007. Positive selection at the protein network periphery: evaluation in terms of structural constraints and cellular context. Proc Natl Acad Sci U S A 104: 20274–20279.

Krylov, D.M., Y.I. Wolf, I.B. Rogozin and E.V. Koonin. 2003. Gene Loss, Protein Sequence Divergence, Gene Dispensability, Expression Level, and Interactivity Are Correlated in Eukaryotic Evolution. Genome Res 13: 2229–2235.

Laghaee, A., C. Malcolm, J. Hallam and P. Ghazal. 2005. Artificial intelligence and robotics in high throughput post-genomics. Drug Discov Today 10(18): 1253–9.

Laportes, D.C., K. Walsh and D.E. Koshland. 1984. The Branch Point Effect. J Biol Chem 259: 14068–14075.

Light, S. and P. Kraulis. 2004. Network analysis of metabolic enzyme evolution in Escherichia coli. BMC Bioinformatics 5: 15.

Livingstone, K. and S. Anderson. 2009. Patterns of variation in the evolution of carotenoid biosynthetic pathway enzymes of higher plants. J Hered 100(6): 754–61. doi: 10.1093/jhered/esp026. Epub 2009 Jun 11. Pub Med PMID: 19520763.

Lu, C., Z. Zhang, L. Leach, M.J. Kearsey and Z.W. Luo. 2007. Impacts of yeast metabolic network structure on enzyme evolution. Genome Biol 8: 407.

Lu, Y. and M.D. Rausher. 2003. Evolutionary rate variation in anthocyanin pathway genes. Mol Biol Evol 20: 1844–53.

Luisi, P., D. Alvarez-Ponce, G.M. Dall'Olio, M. Sikora and J. Bertranpetit. 2012. Network-Level and Population Genetics Analysis of the Insulin/TOR Signal Transduction Pathway Across Human Populations. Mol Biol Evol 29: 1379–1392.

Ma, H.W. and A.P. Zeng. 2003a. The connectivity structure, giant strong component and centrality of metabolic networks. Bioinformatics 19: 1423–1430.

Ma, H.W. and A.P. Zeng. 2003b Reconstruction of metabolic networks from genome data and analysis of their global structure for various organisms. Bioinformatics 19(2): 270–7

Ma, H.W., J. Buer and A.P. Zeng. 2004. Hierarchical structure and modules in the *Escherichia coli* transcriptional regulatory network revealed by a new top-down approach. BMC Bioinformatics 5: 199.

Montanucci, L., H. Laayouni, G.M. Dall'Olio and J. Bertranpetit. 2011. Molecular evolution and network-level analysis of the N-glycosylation metabolic pathway across primates. Mol Biol Evol 28: 813–23.

Montañez, R., M.A. Medina, R.V. Solé and C. Rodríguez-Caso. 2010. When metabolism meets topology: Reconciling metabolite and reaction networks. Bio Essays 32: 246–256.

Nielsen, R. 2005. Molecular Signatures of Natural Selection. Annu Rev Genet 39: 197–218.

Nooren, I.M.A. and J.M. Thornton. 2003. Diversity of protein-protein interactions. EMBO J 22: 3486–3492.

Olson-Manning, C.F., C.R. Lee, M.D. Rausher and T. Mitchell-Olds. 2013. Evolution of flux control in the glucosinolate pathway in Arabidopsis thaliana. Mol Biol Evol 30(1): 14–23.

Papp, B. and S. Oliver. 2005. Genome-wide analysis of the context-dependence of regulatory networks. Genome Biol 6: 206.

Pond, S.L.K., S.D.W. Frost and S.V. Muse. 2005. HyPhy: hypothesis testing using phylogenies. Bioinformatics 21: 676–679.

Pál, C., B. Papp and L.D. Hurst. 2003. Genomic Function: Rate of evolution and gene dispensability. Nature 421: 496–497.

Pál, C., B. Papp and M.J. Lercher. 2006. An integrated view of protein evolution. Nat Rev Genet 7: 337–348.

Pybus, M., G.M. Dall'Olio, P. Luisi, M. Uzkudun, A. Carreño-Torres, P. Pavlidis, H. Laayouni, J. Bertranpetit and J. Engelken. 2013. 1000 Genomes Selection Browser 1.0: a genome browser dedicated to signatures of natural selection in modern humans. Nucl Acids Res doi: 10.1093/nar/gkt1188.

Ramsay, H., L.H. Rieseberg and K. Ritland. 2009. The correlation of evolutionary rate with pathway position in plant terpenoid biosynthesis. Mol Biol Evol 26: 1045–1053.

Rausher, M.D., R.E. Miller and P. Tiffin. 1999. Patterns of evolutionary rate variation among genes of the anthocyanin biosynthetic pathway. Mol Biol Evol 16: 266–274.

Rausher, M.D., Y. Lu and K. Meyer. 2008. Variation in constraint versus positive selection as an explanation for evolutionary rate variation among anthocyanin genes. J Mol Evol 67: 137–144.

Ravasz, E., A.L. Somera, D.A. Mongru, Z.N. Oltvai and A.L. Barabási. 2002. Hierarchical organization of modularity in metabolic networks. Science 297: 1551–1555.

Riley, R.M., W. Jin and G. Gibson. 2003. Contrasting selection pressures on components of the Ras-mediated signal transduction pathway in Drosophila. Mol Ecol 12: 1315–1323.

Rocha, E.P.C. and A. Danchin. 2003. An Analysis of Determinants of Amino Acids Substitution Rates in Bacterial Proteins. Mol Biol Evol 21: 108–116.

Rodriguez-Caso, C., M.A. Medina and R.V. Solé. 2005. Topology, tinkering and evolution of the human transcription factor network. FEBS J 272(24): 6423–34.

Vitkup, D., P. Kharchenko and A. Wagner. 2006. Influence of metabolic network structure and function on enzyme evolution. Genome Biol 7: R39.

Wagner, A. and D.A. Fell. 2001. The small world inside large metabolic networks. Proc Biol Sci 268: 1803–1810.

Wall, D.P., A.E. Hirsh, H.B. Fraser, J. Kumm, G. Giaever, M.B. Eisen and M.W. Feldman. 2005. Functional genomic analysis of the rates of protein evolution. Proc Natl Acad Sci U S A 102: 5483–5488.

Wilson, A.C., S.S. Carlson and T.J. White. 1977. Biochemical evolution. Ann Rev Biochem 46: 573–639.

Wuchty, S. 2004. Evolution and Topology in the Yeast Protein Interaction Network. Genome Res 14: 1310–1314.

Wuchty, S., Z.N. Oltvai and A.L. Barabási. 2003. Evolutionary conservation of motif constituents in the yeast protein interaction network. Nat Genet 35: 176–179.

Yamada, T. and P. Bork. 2009. Evolution of biomolecular networks: lessons from metabolic and protein interactions. Nat Rev Mol Cell Biol 10: 791–803.

Yang, Y.H., F.M. Zhang and S. Ge. 2009. Evolutionary rate patterns of the Gibberellin pathway genes. BMC Evol Biol 9: 206.

Yang, Z. 1997. Likelihood Ratio Tests for Detecting Positive Selection and Application to Primate Lysozyme Evolution. Mol Biol Evol 15: 568–573.

Yang, Z., R. Nielsen, N. Goldman and A. Pedersen. 2000. Codon-Substitution Models for Heterogeneous Selection Pressure at Amino Acid Sites. Genetics 155: 431–49.

Yang, Z. 2007. PAML 4: Phylogenetic Analysis by Maximum Likelihood. Mol Biol Evol 24: 1586–1591.

Yu, H. and M. Gerstein. 2006. Genomic analysis of the hierarchical structure of regulatory networks. Proc Natl Acad Sci U S A 103: 14724–14731.

Zhang, J., R. Nielsen and Z. Yang. 2005. Evaluation of an Improved Branch-Site Likelihood Method for Detecting Positive Selection at the Molecular Level. Mol Biol Evol 22: 2472–2479.

Zhu, X., M. Gerstein and M. Snyder. 2007. Getting connected: analysis and principles of biological networks. Genes Dev 21: 1010–1024.

Chapter 9

Molecular Coevolution: Methods and Applications

Mario A. Fares* and Juan Pablo Labrador

9.1 Introduction

Classic evolutionary theory has focused mainly on the identification of signatures of natural selection acting on single genes or molecules. This narrow view of the action of natural selection clashes with the idea that selection acts on individual organisms but not genes. In addition to this main limitation, a number of criticisms should prevent efforts to identify adaptive evolution in single genes, including that genes encode proteins that are part of large networks of interactions and which fold into complex protein structures involving precise atomic interactions between the amino acids. Therefore, proteins and amino acids rarely evolve in isolation but are part of complex networks of interactions. While most of this book, and indeed others, have devoted entire sections to methods that aim at identifying adaptive events acting on single genes and amino acids, the above

School of Genetics & Microbiology, Dept. of Genetics, University of Dublin, Trinity College, Dublin 2, Dublin, Ireland.
* Corresponding author

stated rationale leads to the conclusion that signatures derived from such methods are often simplistic. To shed light on the evolution of proteins, and indeed gene regulatory circuits, genes should be analyzed in the context of the biological network in which they are embedded. This is better understood if we take into account that each protein encoded by one or more genes contributes with a relative amount to the biological fitness of an individual. This fitness amount is not the result of selection acting on an individual gene but of the relative strength of the interaction of the function encoded by such gene with the remainder of the organismal molecular repertoire. This interaction embarks molecules into a co-adaptation process, in which molecules exercise reciprocal selection upon one another leading to a co-evolutionary dynamic. It follows then that, to understand how living systems work, it is important to discover the evolutionary dependencies among its components as these may point to their functional dependencies.

The term co-evolution was first coined by Ehrlich and Raven to describe the reciprocal evolutionary changes between butterflies and plants (Ehrlich and Raven 1964). Later, Thompson used the word coevolution to refer to the evolutionary dependency between any two populations or species of organisms (Thompson 1994). In his definition, Thompson included the word "reciprocal" evolution, to refer to the bi-directionality of the selective pressures between two interacting populations.

Taking into account that natural selection acts with a relative strength on each of the genes in an organism, we assume that molecular coevolution can be defined as the evolutionary process in which a heritable change in one entity establishes selective pressures that drives changes in another interacting entity. These entities can expand many different levels of complexity including nucleotides, amino acids, proteins and organisms (Fares et al. 2011). In this chapter, we will focus on the molecular coevolution affecting amino acids and proteins. However, a more complex, and likely stronger, coevolutionary relationship exists in the regulatory circuits whose treatment is beyond the scope of this chapter mainly owing to the lack of methods serving such purpose.

9.2 Molecular Coevolution

Once synthesized, proteins sail through a complex energetic folding landscape defined by the atomic interactions among their constituent amino acids. The folding of the protein into a functional conformation requires the interaction between amino acids at the atomic level, so that only compatible amino acid interactions are possible and necessary for protein folding. In essential proteins, which are generally encoded by genes expressed at high levels, amino acid mutations may lead to misfolded proteins and non-specific toxic aggregates dooming the future of the cell (Drummond et al. 2005). Highly expressed, generally essential, protein-coding genes are then under strong constraints and evolve slowly. The amino acids within such proteins are very linked at the evolutionary level so that particular amino acid changes are allowed only if these are accompanied by other compensatory changes at some other sites of the protein (Figure 9.1). This molecular dependencies link amino acids through coadaptation dynamics that generate signatures of coevolution (Codoner and Fares 2008).

Molecular coevolution, as described above (e.g., intra-protein coevolution), relies on the original covariant idea of Fitch and Markowitz, according to which in a particular time point of the evolution of a protein, a once invariable region can undergo changes owing to the relaxed selective

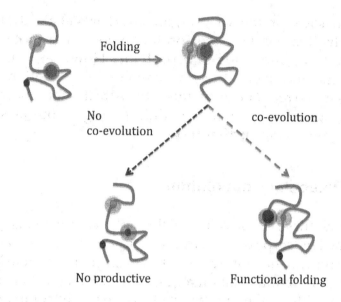

Figure 9.1. Coevolution of amino acids for a productive structure conformation. The red line is a cartoon of an unfolded protein sequence. Blue, green and red colored circles represent amino acid sites of importance for a proper protein folding. The blue and green circles interact through their charges and volume to provide a functional folding. The evolutionary change of the green amino acid to a red one breaks the interaction with the blue amino acid resulting in a faulty structure. This constrains the evolutionary change in the amino acid sequence: a change from the blue amino acid to the green amino acid should be followed by a change to an allowed complementary amino acid in its interacting amino acid site (e.g., one from the green to the blue amino acid). This coadaptation process leads to the coevolution of the blue and green amino acid.

Color image of this figure appears in the color plate section at the end of the book.

constraints promoted by mutations elsewhere in the molecule (Fitch 1971). The concept of coevolution has been applied to many different levels of characters, including those related to morphological variants (Pagel 1994). At the molecular level, including DNA and RNA molecules, coevolution has also been applied to demonstrate that inferences of divergence between DNA sequences are usually underestimates of the real divergence levels when ignoring the correlated evolution of nucleotide sites (Schoniger and von Haeseler 1994). These biases have also been observed when estimating the rates of stem and loops evolution in ribosomal RNA molecules (Rzhetsky 1995).

More recently, the concept of coevolution has been applied to proteins and networks of interacting molecules, the idea being that interacting molecules are likely to evolve together driven by coadaptation dynamics. Indeed, interacting proteins are expected to present protein surface complementarity that when broken through mutations can lead to deleterious effects in cascade in an entire system. It is therefore expected that such changes are not tolerated by natural selection and that mutations in one protein partner at the interaction surface should be followed by mutations in its interacting proteins that are able to restore the integrity of the protein-protein interface. This interaction-centric view has been used in many studies as main assumption to look for new protein-protein interactions using coevolutionary analyses, and thus coevolution can provide an integrative view of a system, an aim that is at the front of biological research (Codoner and Fares 2008; Lovell and Robertson 2010; Fares et al. 2011).

Identification of protein-protein interactions through coevolution analyses has been based on two main parameters that collect the strength and specificity with which proteins interact: the stronger the interaction and the higher the specificity between two proteins the more evolutionarily linked such proteins should be. Based on this assumption, a number of studies have attempted to identify novel protein-protein interactions through correlated phylogenetic profiles in which pairs of proteins were compared for their presence or absence in different

215

organisms: larger correlation in their presence or absence point to their functional or physical interaction (Marcotte et al. 1999). Other studies attempted to identify protein-protein interactions by identifying correlated evolution over the entire protein sequence alignment between two proteins (Goh et al. 2000; Goh and Cohen 2002; Valencia and Pazos 2003; Pazos et al. 2005; Hakes et al. 2007; Juan et al. 2008; de Juan et al. 2013). Finally, a number of other studies have used site-specific correlated signatures of evolution between proteins as indication of protein-protein interactions (Gloor et al. 2005; Martin et al. 2005; Mintseris and Weng 2005; Fares and Travers 2006; Madaoui and Guerois 2008). Finally, molecular coevolution within proteins has also been used to identify the interactions between amino acids that may be key to protein structure. Indeed, amino acids within proteins coevolve to some extent to maintain the structural integrity of the protein. Such coevolution may also be driven by complex atomic interactions during the protein folding process. Therefore, proteins that neighbor each other in the protein structure are likely to be reciprocally constrained, exercising strong selective forces upon one another (Altschuh et al. 1987; Clarke 1995; Atchley et al. 2000; Gloor et al. 2005; Yeang and Haussler 2007; Codoner and Fares 2008; Halabi et al. 2009; Lovell and Robertson 2010). Based on the rationale above, coevolution is thus expected in all proteins. However, because signatures of molecular coevolution comprise undetermined noise caused by historical and stochastic amino acids covariation (Atchley et al. 2000), functional and structural coevolution could be disentangled from stochastic and historical covariation in only a number of proteins known to be essential for organism's survival. The list of such proteins include sheat-shock proteins (Fares and Travers 2006; Travers and Fares 2007; Ruiz-Gonzalez and

216

Fares 2013), HIV envelope proteins (Travers et al. 2007), Cytochrome C oxidase (Wang and Pollock 2007) and many other proteins (Gloor et al. 2005). Increasing the capacity, specificity and sensitivity, of methods to identify functional coevolution will help unearthing the complex coadaptation dynamics within and between proteins that govern proteins folding or molecular interactions, although this remains a future objective.

9.3 Methods of Molecular Coevolution

Broadly speaking, there are two groups of coevolution methods, those aiming at identifying coevolution between proteins and those that identify evolutionary dependencies among amino acid sites within or between proteins. In this chapter is not our aim to provide a detailed description of all these methods, for which we propose to read a recent review (de Juan et al. 2013). We instead will focus on the methodological and pragmatic aspects of methods devoted to the identification of coevolution at the residue level. Within this category, we can identify parametric and non-parametric methods, many of which are based on principles of molecular phylogenetics. As for methods that identify coevolution between proteins, most identify correlation in the phylogenies of two proteins using phylogenetic distance based matrices, with those proteins showing mirroring phylogenetic profiles being candidates for coevolving at the molecular level (Figure 9.2) (Gobel et al. 1994; Pazos et al. 1997a; Pazos et al. 1997b; Goh et al. 2000; Pazos and Valencia 2001; Goh and Cohen 2002; Pazos and Valencia 2002; Hamilton et al. 2004; Kim et al. 2004; Pazos et al. 2005; Jothi et al. 2006; Waddell et al. 2007).

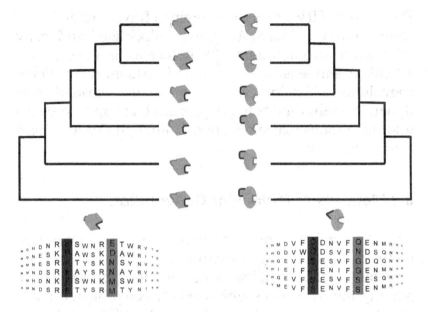

Figure 9.2. Coevolution and tree mirroring between two interacting proteins. The red proteins surface regions are those that mediate protein interactions (red amino acid sites in the multiple sequence alignment), while the green regions are not involved in protein interfaces (red amino acid sites in the multiple sequence alignment). Changes in the geometry of the red region in one protein constrain the geometry of the red region of its interacting protein partner, while this is not the case for the green regions. These reciprocal constrain lead to mirroring phylogenetic inferences when using the two interacting proteins.

Color image of this figure appears in the color plate section at the end of the book.

9.3.1 Non-parametric Methods of Molecular Coevolution

The most representative non-parametric methods of molecular coevolution are those based on measuring the Mutual Information Content (MIC) for pairs of amino acid sites. These methods use the information theory (Kullback

1959) to find and quantify the variability at amino acid sites within a multiple sequence alignment (MSA). In brief, this method calculates the Shanon entropy for each amino acid site in the MSA (H_i). This Shanon entropy is measured as the probability of each of the 20 amino acid states (A, S, L,...) to be at site i of the multiple sequence alignment, calculated as:

$$H(i) = - \sum_{x=A,S,L,...} P(x_i) log P(x_i)$$

The probability of co-occurrence of two particular amino acid states (x and y) at two positions (i and j), respectively, can also be calculated as:

$$H(i,j) = - \sum_{x_i, y_j} P(x_i, y_j) log P(x_i, y_j)$$

From these quantities, one can calculate how much information is mutual, or shared, between two amino acid sites (i and j) by relating the amount of information in each of the sites to the amount of shared information by the two amino acid sites as:

$$MI = H(i) + H(j) - H(i,j)$$

Notice that $H(i)$ and $H(j)$ are both negative values. Under this definition, if the two sites being compared (i and j) evolve independently, then the amount of variability in each site should be equivalent to the amount of covariability for the two amino acid sites and MI should yield a value not significantly different from 0. Alternatively, if the two sites evolve together, then $H(i,j)$ should provide a larger negative value than $H(i)$ and $H(j)$, with MI yielding a positive value significantly greater than 0. Therefore, MI is a value ranging between 0 and a positive value whose magnitude is proportional to the quantity of coevolution between two amino acid sites. Methods relying on measurement of mutual information content are subjected to large false positive values, mainly dependent on the number of sequences in the MSA as well as the level of conservation (Fodor and Aldrich 2004; Martin et al. 2005; Fares and Travers 2006). A number of additions to the previous calculations have been conducted by other studies to increase the sensitivity and specificity of coevolution methods relying on MI. For example, some studies used relevant biological information to reduce the false positives rates in MI coevolution analysis, such as the correlation in the physicochemical properties between amino acid pairs with significant values of MI (Martin et al. 2005; Codoner et al. 2008). Other methods rely on mathematical corrections to increase the accuracy of *MI* calculations and remove stochastic and phylogenetic covariation (Tillier et al. 2006; Dunn et al. 2008; Rodionov et al. 2011).

Coevolution analysis using MI can be carried out using one of the available servers for such purpose (http:// coevolution.gersteinlab.org/coevolution/index.jsp). Figure 9.3 shows the set of parameters that need to be specified in order to run this server. A multiple sequence alignment and a phylogenetic tree in Newick format can be uploaded into the server. A number of methods to identify correlations between amino acid sites can be used, of which we will focus on the analysis of Mutual Information. In addition to the coevolution analysis, scores can be used to plot significant pairs of amino acid sites in the three-dimensional structure, if available, which can be directly downloaded from the structure database at the National Centre of Bioinformatics.

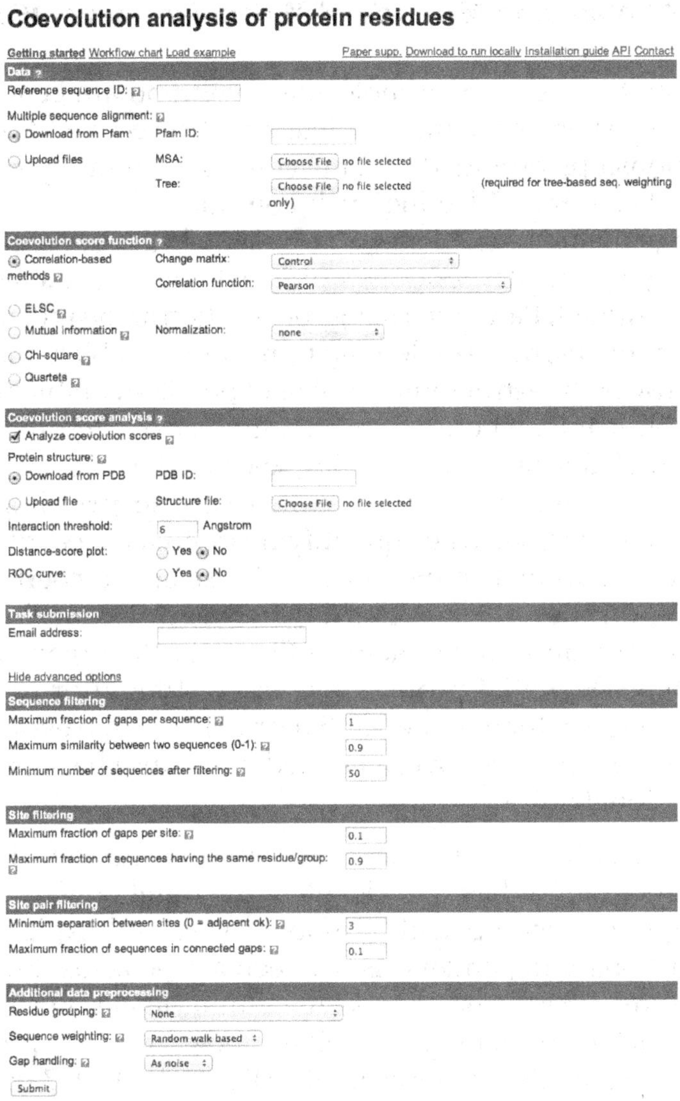

Coevolution analysis of protein residues

Getting started Workflow chart Load example   Paper supp. Download to run locally Installation guide API Contact

Data ?

Reference sequence ID: ☐ [   ]

Multiple sequence alignment: ☐
- ⦿ Download from Pfam   Pfam ID: [   ]
- ○ Upload files   MSA: [Choose File] no file selected
  Tree: [Choose File] no file selected (required for tree-based seq. weighting only)

Coevolution score function ?

- ⦿ Correlation-based methods ☐   Change matrix: [Control ▾]   Correlation function: [Pearson ▾]
- ○ ELSC ☐
- ○ Mutual information ☐   Normalization: [none ▾]
- ○ Chi-square ☐
- ○ Quartets ☐

Coevolution score analysis ?

☑ Analyze coevolution scores ☐

Protein structure: ☐
- ⦿ Download from PDB   PDB ID: [   ]
- ○ Upload file   Structure file: [Choose File] no file selected

Interaction threshold: [6] Angstrom
Distance-score plot: ○ Yes ⦿ No
ROC curve: ○ Yes ⦿ No

Task submission

Email address: [   ]

Hide advanced options

Sequence filtering

Maximum fraction of gaps per sequence: ☐ [1]
Maximum similarity between two sequences (0-1): ☐ [0.9]
Minimum number of sequences after filtering: ☐ [50]

Site filtering

Maximum fraction of gaps per site: ☐ [0.1]
Maximum fraction of sequences having the same residue/group: ☐ [0.9]

Site pair filtering

Minimum separation between sites (0 = adjacent ok): ☐ [3]
Maximum fraction of sequences in connected gaps: ☐ [0.1]

Additional data preprocessing

Residue grouping: ☐ [None ▾]
Sequence weighting: ☐ [Random walk based ▾]
Gap handling: ☐ [As noise ▾]

[Submit]

Figure 9.3. Analysis of Mutual Information content between amino acid sites in a multiple protein sequence alignment (MSA). A number of parameters need to be specified by the user, including the correlation function, the proteins structure PDB file, and Fraction of allowed gaps in the amino acid site column of the MSA. Results are sent to the user's email address.

221

9.3.2 Parametric Methods of Molecular Coevolution

Parametric methods of molecular evolution are generally more specific or accurate in detecting true coevolutionary relationships provided that the conditions underlying the models are met. Meeting such conditions, however rare, may provide valuable results in which the coevolution between particular residues within or between proteins can be quantified. Parameterizing coevolution at the molecular level remains, nevertheless, an unresolved problem mainly because of the astronomic number of possible combinations of atomic interactions between amino acids, and the lack of knowledge on the underlying selective forces governing such interactions. In addition to these problems, a fundamental issue when quantifying molecular coevolution through parametric methods is the selective coefficient acting on the protein—that is, the relative contribution of the protein to the biological fitness of the organism as a whole. Essential proteins are more straightforward to parameterize than non-essential proteins, as the evolution of the former is less overwhelmed by neutral background mutational noise.

Notwithstanding the problems mentioned above, enormous effort has been devoted by a number of research groups to come up with a model that could explain the evolutionary dependencies between amino acids. Indeed, some researchers have used maximum-likelihood based approaches to identify coevolving positions in a protein (Pollock et al. 1999; Choi et al. 2005; Pei et al. 2006), Bayesian framework (Dimmic et al. 2005), molecular phylogenetics (Fukami-Kobayashi et al. 2002), or correlated corrected-divergence matrices (Fares and McNally 2006; Fares and Travers 2006). Of these methods, that of Fares and Travers has shown to improve previous predictions of molecular coevolution, and thus we will focus on the methodological and technical aspects of this method.

9.3.3 CAPS: Coevolution Analysis using Protein Sequences

Fares and Travers developed a model (hereafter called the CAPS model) of coevolution in which the patterns of amino acid replacements between any pair of amino acids, within or between proteins, are compared. The key calculation making this method accurate is the correction of the amino acid divergence between two sequences at a particular amino acid site by the likelihood of the amino acid transition at that site. Let's assume an alignment as the one shown in Figure 9.4, that comprises a number of eight aligned orthologous amino acid sequences (Seq1 to Seq8).

For each of the amino acid sites in that alignment, we can calculate the likelihood of the transition between the amino acid states for any two sequences at that site using **BLO**ck **SU**bstitution **M**atrices (BLOSUM, (Henikoff and Henikoff 1992)). BLOSUM are empirical matrices that evaluate the probability of observing an amino acid transition, with negative, 0 and positive values meaning that the transition is very unlikely, expected, or highly expected, respectively. Using this blosum values (here symbolized as $(B_{ek})_i$, indicating the transition between the amino acid states e and k for site i in the alignment), we can estimate the amount of evolution for a particular pair of sequences as:

$$\theta_i = (B_{ek}t^{-1})$$

Here, t stands for the total distance between the two sequences being compared using either amino acid distances or synonymous nucleotides differences. Therefore, we

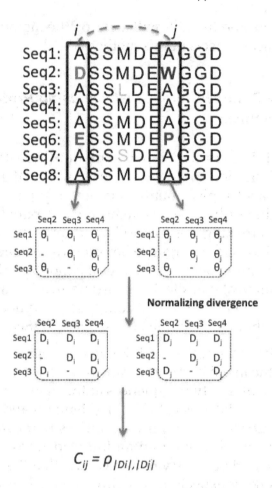

Figure 9.4. CAPS theoretical workflow. A multiple sequence alignment, either based on amino acids or codon sequences, is used to identify pairs of sites with signatures of coevolution. In the example of the figure, amino acid sites *i* and *j* are tested for coevolution (C_{ij}) by measuring the correlation in their divergence patterns across an alignment of 8 sequences. CAPS normalizes the divergence patterns by the mean pairwise amino acid sequence distance to account for the divergence time between sequences as well as by the level of divergence for that site in the alignment.

Color image of this figure appears in the color plate section at the end of the book.

effectively correct the transition by the relative divergence time between two sequences: the larger the time the greater is the probability for a radical amino acid transition (e.g., from Ala to Asp) to occur. It follows then that big amino acid transitions at short evolutionary divergences have a large weight in the calculation, and vice versa. Because some amino acid sites evolve quicker than others, hence their θ values is hardly comparable, we normalize the transition values by subtracting the mean transition for that site across the alignment from each of the individual transitions and squaring the result:

$$D_i = (\theta_i - \bar{\theta}_i)^2$$

This transformation yields a normalized transition matrix of amino acid substitutions for each of the amino acid sites that can now be correlated using standard methods (Figure 9.5) (see details of the model in (Fares and Travers 2006)).

In addition to the mathematical transformation of CAPS model, recent improvements in CAPS allow increasing the sensitivity of the method by including the phylogenetic relationships between the sequences of the MSA and comparing the correlation coefficients with a null distribution of these coefficients. The null distribution has also improved in comparison with existing methods, as this is drawn from simulated sequence alignments using the same amino acid composition, phylogenetic relationships, and divergence levels as the real sequence alignment. This simulation procedure allows accounting for biases which are difficult to correct, mainly the phylogeny asymmetries or long-branch attraction effects and recombination, factors that can significantly influence the detection of coevolving pairs of amino acids.

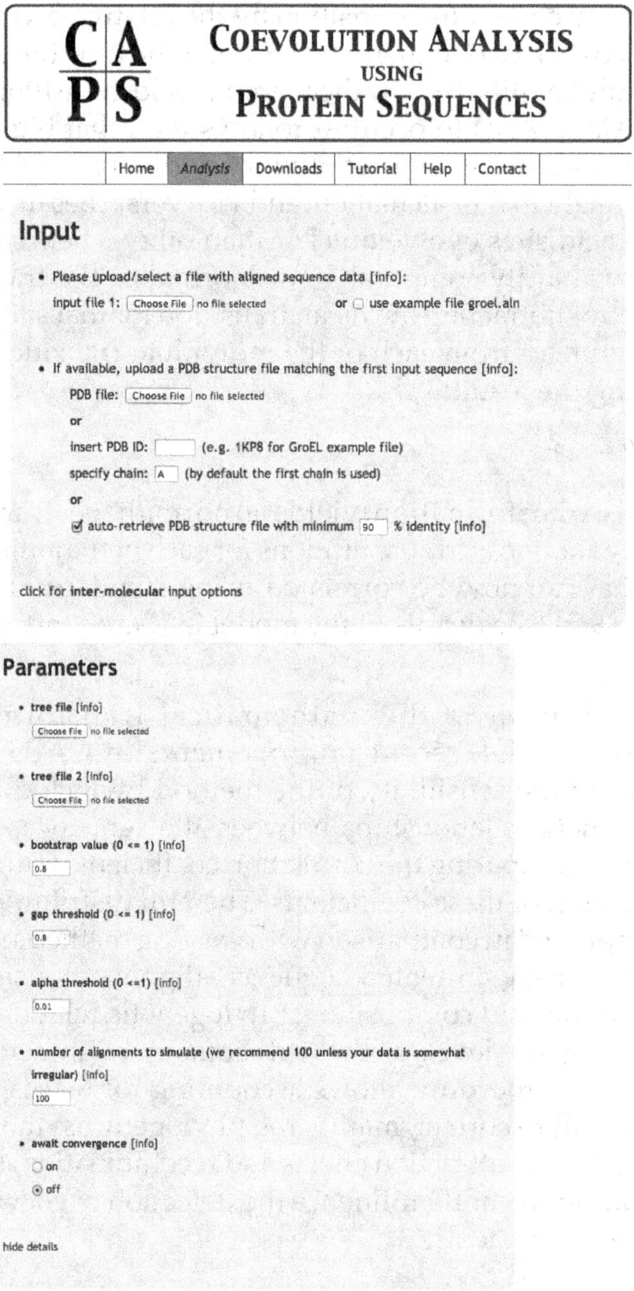

C|A P|S COEVOLUTION ANALYSIS
USING
PROTEIN SEQUENCES

Home | Analysis | Downloads | Tutorial | Help | Contact

Input

- Please upload/select a file with aligned sequence data [info]:

  input file 1: [Choose File] no file selected   or ○ use example file groel.aln

- If available, upload a PDB structure file matching the first input sequence [info]:

  PDB file: [Choose File] no file selected

  or

  insert PDB ID: [____] (e.g. 1KP8 for GroEL example file)

  specify chain: [A] (by default the first chain is used)

  or

  ☑ auto-retrieve PDB structure file with minimum [90] % identity [info]

click for **inter-molecular** input options

Parameters

- **tree file** [info]
  [Choose File] no file selected

- **tree file 2** [info]
  [Choose File] no file selected

- **bootstrap value (0 <= 1)** [info]
  [0.8]

- **gap threshold (0 <= 1)** [info]
  [0.8]

- **alpha threshold (0 <=1)** [info]
  [0.01]

- **number of alignments to simulate** (we recommend 100 unless your data is somewhat **irregular**) [info]
  [100]

- **await convergence** [info]
  ○ on
  ◉ off

hide details

Figure 9.5. contd....

226

To illustrate an example using this method, we have ran CAPS model in an alignment of amino acid sequences belonging to the heat-shock protein GroEL from a diverse set of bacteria. GroEL is an essential protein in bacteria and eukaryotic organelles under normal physiological conditions and heat stress. CAPS program, that comprises the model described above (Fares and McNally 2006), can be run online in (http://caps.tcd.ie) which provides a user-friendly management of the program for running coevolution using protein-coding nucleotide sequences or amino acid sequences. Few parameters need to be explained before running our example using the CAPS webserver (Figure 9.5). First, the user should upload a multiple sequence alignment in the system, and optionally the structure flat file (if available) containing the atomic coordinates in the space for the amino acids, and a phylogenetic tree in Newick format. In addition to this, the method has been equipped with a non-parametric bootstrap analysis, in which for each pair of identified coevolving amino acids, the analysis is repeated for a 1000 sub-sequences to identify those pairs that coevolve across the entire phylogeny (for example, they show bootstrap values equal or larger than 70%) and those that coevolve in a small set of aligned sequences (i.e., bootstrap values lower than 70%). The amount of tolerated gapped sequences at a particular amino acid position can also be specified. Finally, the type I error (e.g., the proportion

Figure 9.5. contd.

Figure 9.5. CAPS webserver parameters. A number of parameters can be set to run CAPS online, including the multiple sequence alignment files, the phylogenetic tree and the protein structure PDB file. Non-parametric bootstrapping can also be performed to identify transient and permanent coevolving pairs of sites. Gaps are allowed in a proportion of the sequences as specified by the user. Results generated are sent by email to the user.

of cases in the tails of the distribution) can be set, while it is generally recommended to use $\alpha = 0.01$.

The results generated when running CAPS on the GroEL alignment yield 15 groups of coevolution. Groups of coevolution are defined as sets of amino acids where each coevolves with every other amino acid within the same set (Figure 9.6a represents a sample of the relationships between the groups of coevolution).

The output generated also provides a table with columns 2 to 5 presenting information on the group number, the number of residues in the group, the number of residues with three-dimensional structure information, and the mean distance and sites involved, respectively (Figure 9.6b). Three-dimensional representation of coevolving amino acids of group 1 (Figure 9.7) allows inspecting the spatial relationship between amino acid sites, many of which fall close of one another in the structure, indicating that coevolution may be the result of coadaptation dynamics. In terms of the estimates of parameters for that particular group, these are provided in an additional table in an output file that shows the correlation coefficient for each pair of sites, their bootstrap support and three-dimensional distance taking the most proximal carbon atoms.

A fundamental step to better interpret the results originated using this method is finding functional information on the sites under coevolution. It is particularly important to realize that functionally coevolving amino acid sites must necessarily be linked to adaptive processes that allow the fixation of changes improving function or structure. In the analysis presented here, most if not all amino acids involved in a coevolutionary relationship have been reported to have a functional role in the binding of unfolded proteins by GroEL, an essential step in protein

(a)
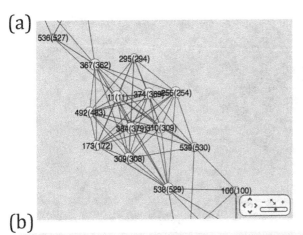

(b)

Listing of groups with residues covered by 3-D structure (click on headers to sort):

o	Group [?]	Sites [?]	3-D sites [?]	Residue indices [?]
⊙ 1	14	10	distance:, mean, 11, 36.0558, **172**, **254**, **294**, **308**, **309**, **362**, **369**, **379**, **483**, 529
○ 2	14	9	distance:, mean, 11, 37.6969, **254**, **294**, **308**, **309**, **362**, **369**, **379**, **483**, **529**, **530**
○ 3	9	5	distance:, mean, 11, 42.6869, **362**, **369**, **379**, **483**, **527**
○ 4	8	4	distance:, mean, 41.944, **48**, **100**, **133**, **239**, **530**

Figure 9.6. Output generated running CAPS on the heat-shock protein GroEL from gamma-proteobacteria. (a) A coevolution network generated taking into account the coevolutionary links between amino acids in the protein. Numbers without brackets indicate the position in the multiple sequence alignment, while those within brackets refer to the position in a reference sequence once gaps have been removed. (b) Table of coevolution groups. The mean Euclidean structural distance in Angstroms for all pairwise amino acid comparisons within the group is given. In bold are those amino acid sites that have been crystallized. For example, in group 1, there is spatial information on the GroEL structure for all sites with the exception of site 529.

Color image of this figure appears in the color plate section at the end of the book.

229

Figure 9.7. The protein structure of GroEL (code: 1PCQ.pdb). Sites within the same group of coevolution are labeled with the same color.

Color image of this figure appears in the color plate section at the end of the book.

folding in the cell. For example, the amino acid sites that form part of the group highlighted in Figure 9.6 (sites 254, 294, 308, and 309) are directly involved in the binding of unfolded proteins by the hydrophobic surface in the apical domain of GroEL subunits. These four amino acids define the binding surface and their coevolution is definitively pointing to their functional link.

9.4 Coevolution in Biological Networks: The Case of Ligand-receptor Interactions

Evolution of complexity in particular in the immune system and the nervous system has been driven by cell-cell communication. This communication is mediated through receptors expressed in membranes and ligands that can signal in a long-range or short-range mode depending on whether they diffuse away from the producing cell or remain attached to it respectively. Molecular changes in one molecule should result in variation of its interacting partner to preserve ligand-receptor specificity across evolution. Additionally, increasing complexity has been paralleled by duplications within receptor and ligand families whose interactions have diversified through the generation of novel specificities (Moyle et al. 1994; Fares et al. 2011). Thus, cell surface and secreted molecules present a very clear example where coevolution could be exploited to identify novel interactions. However, these types of interactions have been harder to identify with high throughput molecular technologies such as yeast two-hybrid or affinity purification followed by mass spectrometry. Molecular coevolution methods should provide a complementary approach to the identification of such interactions.

Indeed, methods exploring the correlation in the phylogenies of families of proteins as discussed in section 9.3 have been successful to reduce the search space for the identification of binding partners among ligands and receptors and identifying novel interactions. For example, Goh et al. (Goh et al. 2000), used the chemokine family of protein ligands and their G-protein coupled receptors as a model to demonstrate that both families had coevolved in such a way that each subgroup of chemokines would interact with a specific subgroup of receptors. Based on

231

this approach they predicted a novel interaction for the H174/IP-9 chemokine with the CXCR3 receptor, which was later confirmedexperimentally (Tensen et al. 1999). Further analyses within the TGFβ/TGFβ receptor subfamilies predicted the uncharacterized interactions of GDF6 and GDF7 with the BMP receptors 1A and 1B as well as those of GDF8 and GDF11 with type 2 activin receptors (Goh and Cohen 2002). These interactions have been also subsequently confirmed with molecular methods (Mazerbourg et al. 2005; Sako et al. 2010). While methods based in correlated evolution of protein families have successfully identified novel interactions, as described above, they are limited to the protein families studied rather than providing a global view of the extracellular interactome (ECI).

A recent molecular approach has been developed by Ökzan et al. to specifically study the ECI in *Drosophila melanogaster* using ectodomains of membrane and secreted proteins. They characterized interactions among 202 proteins belonging to the immunoglobulin, fibronectin type III and leucin rich repeat families in *Drosophila*. The novel methodology used extracellular domains arrayed on a solid matrix and further assayed their interaction with soluble alkaline phosphates fusion of the same ectodomains. In total over 20,000 unique pairwise putative interactions were tested and 83 previously unknown interactions were described (Ozkan et al. 2013). Although the authors noted some understandable limitations of their method, in particular their false negative rate due to low protein expression for some of their constructs, the collection of ligand-receptor interactions generated provides an extremely high quality database of the partial *Drosophila* ECI previously unavailable. This resource provides the essential framework for computational methods to study ECIs of multicellular organisms, and in particular the insect clade, as it provides an excellent calibration tool. In addition, *Drosophila* represents a particularly useful species for coevolutionary studies as 12 species have been sequenced (Drosophila 12 Genomes et al. 2007) allowing for enough orthologous genes (coding for the same protein in different species) and evolutionary signal (amino acid variation) when performing coevolutionary analyses (Codoner et al. 2008).

Most of the methods aiming at identifying molecular interactions through coevolution rely on computationally expensive algorithms. CAPS v2.0 is a newly emerging program that implements fast algorithms devoted to the high throughput analyses and which allow undertaking the challenge to uncover signatures of strong coevolution among biologically interacting molecules.

References

Altschuh, D., A.M. Lesk, A.C. Bloomer and A. Klug. 1987. Correlation of co-ordinated amino acid substitutions with function in viruses related to tobacco mosaic virus. Journal of Molecular Biology 193: 693–707.

Atchley, W.R., K.R. Wollenberg, W.M. Fitch, W. Terhalle and A.W. Dress. 2000. Correlations among amino acid sites in bHLH protein domains: an information theoretic analysis. Molecular Biology and Evolution 17: 164–178.

Choi, S.S., W. Li and B.T. Lahn. 2005. Robust signals of coevolution of interacting residues in mammalian proteomes identified by phylogeny-aided structural analysis. Nature Genetics 37: 1367–1371. doi: 10.1038/ng1685.

Clarke, N.D. 1995. Covariation of residues in the homeodomain sequence family. Protein Science: A Publication of the Protein Society 4: 2269–2278. doi: 10.1002/pro.5560041104.

Codoner, F.M. and M.A. Fares. 2008. Why should we care about molecular coevolution? Evolutionary Bioinformatics Online 4: 29–38.

Codoner, F.M., S. O'Dea and M.A. Fares. 2008. Reducing the false positive rate in the non-parametric analysis of molecular coevolution. BMC Evolutionary Biology 8: 106. doi: 10.1186/1471-2148-8-106.

de Juan, D., F. Pazos and A. Valencia. 2013. Emerging methods in protein co-evolution. Nature Reviews Genetics 14: 249–261. doi: 10.1038/nrg3414.

Dimmic, M.W., M.J. Hubisz, C.D. Bustamante and R. Nielsen. 2005. Detecting coevolving amino acid sites using Bayesian mutational mapping. Bioinformatics 21 Suppl. 1: i126–135. doi: 10.1093/bioinformatics/bti1032.

Drosophila 12 Genomes C, A.G. Clark, M.B. Eisen et al. 2007. Evolution of genes and genomes on the Drosophila phylogeny. Nature 450: 203–218. doi: 10.1038/nature06341.

Drummond, D.A., J.D. Bloom, C. Adami, C.O. Wilke and F.H. Arnold. 2005. Why highly expressed proteins evolve slowly. Proceedings of the National Academy of Sciences of the United States of America 102: 14338–14343. doi: 10.1073/pnas.0504070102.

Dunn, S.D., L.M. Wahl and G.B. Gloor. 2008. Mutual information without the influence of phylogeny or entropy dramatically improves residue contact prediction. Bioinformatics 24: 333–340. doi: 10.1093/bioinformatics/btm604.

Ehrlich, P.R. and P.H. Raven. 1964. Butterflies and plants: a study in coevolution. Evolution 18: 23.

Fares, M.A. and D. McNally. 2006. CAPS: coevolution analysis using protein sequences. Bioinformatics 22: 2821–2822. doi: 10.1093/bioinformatics/btl493.

Fares, M.A., M.X. Ruiz-Gonzalez and J.P. Labrador. 2011. Protein coadaptation and the design of novel approaches to identify protein-protein interactions. IUBMB Life 63: 264–271. doi: 10.1002/iub.455.

Fares, M.A. and S.A. Travers. 2006. A novel method for detecting intramolecular coevolution: adding a further dimension to selective constraints analyses. Genetics 173: 9–23. doi: 10.1534/genetics.105.053249.

Fitch, W.M. 1971. Rate of change of concomitantly variable codons. Journal of Molecular Evolution 1: 84–96.

Fodor, A.A. and R.W. Aldrich. 2004. Influence of conservation on calculations of amino acid covariance in multiple sequence alignments. Proteins 56: 211–221. doi: 10.1002/prot.20098.

Fukami-Kobayashi, K., D.R. Schreiber and S.A. Benner. 2002. Detecting compensatory covariation signals in protein evolution using reconstructed ancestral sequences. Journal of Molecular Biology 319: 729–743. doi: 10.1016/S0022-2836(02)00239-5.

Gloor, G.B., L.C. Martin, L.M. Wahl and S.D. Dunn. 2005. Mutual information in protein multiple sequence alignments reveals two classes of coevolving positions. Biochemistry 44: 7156–7165. doi: 10.1021/bi050293e.

Gobel, U., C. Sander, R. Schneider and A. Valencia. 1994. Correlated mutations and residue contacts in proteins. Proteins 18: 309–317. doi: 10.1002/prot.340180402.

Goh, C.S., A.A. Bogan, M. Joachimiak, D. Walther and F.E. Cohen. 2000. Co-evolution of proteins with their interaction partners. Journal of Molecular Biology 299: 283–293. doi: 10.1006/jmbi.2000.3732.

ROF

Goh, C.S. and F.E. Cohen. 2002. Co-evolutionary analysis reveals insights into protein-protein interactions. Journal of Molecular Biology 324: 177–192.

Hakes, L., S.C. Lovell, S.G. Oliver and D.L. Robertson. 2007. Specificity in protein interactions and its relationship with sequence diversity and coevolution. Proceedings of the National Academy of Sciences of the United States of America 104: 7999–8004. doi: 10.1073/pnas.0609962104.

Halabi, N., O. Rivoire, S. Leibler and R. Ranganathan. 2009. Protein sectors: evolutionary units of three-dimensional structure. Cell 138: 774–786. doi: 10.1016/j.cell.2009.07.038.

Hamilton, N., K. Burrage, M.A. Ragan and T. Huber. 2004. Protein contact prediction using patterns of correlation. Proteins 56: 679–684. doi: 10.1002/prot.20160.

Henikoff, S. and J.G. Henikoff. 1992. Amino acid substitution matrices from protein blocks. Proceedings of the National Academy of Sciences of the United States of America 89: 10915–10919.

Jothi, R., P.F. Cherukuri, A. Tasneem and T.M. Przytycka. 2006. Co-evolutionary analysis of domains in interacting proteins reveals insights into domain-domain interactions mediating protein-protein interactions. Journal of Molecular Biology 362: 861–875. doi: 10.1016/j.jmb.2006.07.072.

Juan, D., F. Pazos and A. Valencia. 2008. High-confidence prediction of global interactomes based on genome-wide coevolutionary networks. Proceedings of the National Academy of Sciences of the United States of America 105: 934–939. doi: 10.1073/pnas.0709671105.

Kim, W.K., D.M. Bolser and J.H. Park. 2004. Large-scale co-evolution analysis of protein structural interlogues using the global protein structural interactome map (PSIMAP). Bioinformatics 20: 1138–1150. doi: 10.1093/bioinformatics/bth053.

Kullback, S. 1959. Information theory and statistics. New York: Willey.

Lovell, S.C. and D.L. Robertson. 2010. An integrated view of molecular coevolution in protein-protein interactions. Molecular Biology and Evolution 27: 2567–2575. doi: 10.1093/molbev/msq144.

Madaoui, H. and R. Guerois. 2008. Coevolution at protein complex interfaces can be detected by the complementarity trace with important impact for predictive docking. Proceedings of the National Academy of Sciences of the United States of America 105: 7708–7713. doi: 10.1073/pnas.0707032105.

Marcotte, E.M., M. Pellegrini, H.L. Ng, D.W. Rice, T.O. Yeates and D. Eisenberg. 1999. Detecting protein function and protein-protein interactions from genome sequences. Science 285: 751–753.

Martin, L.C., G.B. Gloor, S.D. Dunn and L.M. Wahl. 2005. Using information theory to search for co-evolving residues in proteins. Bioinformatics 21: 4116–4124. doi: 10.1093/bioinformatics/bti671.

Mazerbourg, S., K. Sangkuhl, C.W. Luo, S. Sudo, C. Klein and A.J. Hsueh. 2005. Identification of receptors and signaling pathways for orphan bone morphogenetic protein/growth differentiation factor ligands based on genomic analyses. The Journal of Biological Chemistry 280: 32122–32132. doi: 10.1074/jbc.M504629200.

Mintseris, J. and Z. Weng. 2005. Structure, function, and evolution of transient and obligate protein-protein interactions. Proceedings of the National Academy of Sciences of the United States of America 102: 10930–10935. doi: 10.1073/pnas.0502667102.

Moyle, W.R., R.K. Campbell, R.V. Myers, M.P. Bernard, Y. Han and X. Wang. 1994. Co-evolution of ligand-receptor pairs. Nature 368: 251–255. doi: 10.1038/368251a0.

Ozkan, E., R.A. Carrillo, C.L. Eastman, R. Weiszmann, D. Waghray, K.G. Johnson, K. Zinn, S.E. Celniker and K.C. Garcia. 2013. An extracellular interactome of immunoglobulin and LRR proteins reveals receptor-ligand networks. Cell 154: 228–239. doi: 10.1016/j.cell.2013.06.006.

Pagel, M. 1994. Detecting correlated evolution on phylogenies: A general method for the comparative analysis of discrete characters. Proceedings of the Royal Society of Biological Sciences Ser B 255: 9.

Pazos, F., M. Helmer-Citterich, G. Ausiello and A. Valencia. 1997a. Correlated mutations contain information about protein-protein interaction. Journal of Molecular Biology 271: 511–523. doi: 10.1006/jmbi.1997.1198.

Pazos, F., O. Olmea and A. Valencia. 1997b. A graphical interface for correlated mutations and other protein structure prediction methods. Computer Applications in the Biosciences: CABIOS 13: 319–321.

Pazos, F., J.A. Ranea, D. Juan and M.J. Sternberg. 2005. Assessing protein co-evolution in the context of the tree of life assists in the prediction of the interactome. Journal of Molecular Biology 352: 1002–1015. doi: 10.1016/j.jmb.2005.07.005.

Pazos, F. and A. Valencia. 2001. Similarity of phylogenetic trees as indicator of protein-protein interaction. Protein Engineering 14: 609–614.

Pazos, F. and A. Valencia. 2002. *In silico* two-hybrid system for the selection of physically interacting protein pairs. Proteins 47: 219–227.

Pei, J., W. Cai, L.N. Kinch and N.V. Grishin. 2006. Prediction of functional specificity determinants from protein sequences using log-likelihood ratios. Bioinformatics 22: 164–171. doi: 10.1093/bioinformatics/bti766.

Pollock, D.D., W.R. Taylor and N. Goldman. 1999. Coevolving protein residues: maximum likelihood identification and relationship to structure. Journal of Molecular Biology 287: 187–198. doi: 10.1006/jmbi.1998.2601.

Rodionov, A., A. Bezginov, J. Rose and E.R. Tillier. 2011. A new, fast algorithm for detecting protein coevolution using maximum compatible cliques. Algorithms for Molecular Biology: AMB 6: 17. doi: 10.1186/1748-7188-6-17.

Ruiz-Gonzalez, M.X. and M.A. Fares. 2013. Coevolution analyses illuminate the dependencies between amino acid sites in the chaperonin system GroES-L. BMC Evolutionary Biology 13: 156. doi: 10.1186/1471-2148-13-156.

Rzhetsky, A. 1995. Estimating substitution rates in ribosomal RNA genes. Genetics 141: 771–783.

Sako, D., A.V. Grinberg, J. Liu et al. 2010. Characterization of the ligand binding functionality of the extracellular domain of activin receptor type IIb. The Journal of Biological Chemistry 285: 21037–21048. doi: 10.1074/jbc.M110.114959.

Schoniger, M. and A. von Haeseler. 1994. A stochastic model for the evolution of autocorrelated DNA sequences. Molecular Phylogenetics and Evolution 3: 240–247. doi: 10.1006/mpev.1994.1026.

Tensen, C.P., J. Flier, E.M. Van Der Raaij-Helmer, S. Sampat-Sardjoepersad, R.C. Van Der Schors, R. Leurs, R.J. Scheper, D.M. Boorsma and R. Willemze. 1999. Human IP-9: A keratinocyte-derived high affinity CXC-chemokine ligand for the IP-10/Mig receptor (CXCR3). The Journal of Investigative Dermatology 112: 716–722. doi: 10.1046/j.1523-1747.1999.00581.x.

Thompson, J.N. 1994. The coevolutionary Process. Chicago: University of Chicago Press.

Tillier, E.R., L. Biro, G. Li and D. Tillo. 2006. Codep: maximizing co-evolutionary interdependencies to discover interacting proteins. Proteins 63: 822–831. doi: 10.1002/prot.20948.

Travers, S.A. and M.A. Fares. 2007. Functional coevolutionary networks of the Hsp70-Hop-Hsp90 system revealed through computational analyses. Molecular Biology and Evolution 24: 1032–1044. doi: 10.1093/molbev/msm022.

Travers, S.A., D.C. Tully, G.P. McCormack and M.A. Fares. 2007. A study of the coevolutionary patterns operating within the env gene of the HIV-1 group M subtypes. Molecular Biology and Evolution 24: 2787–2801. doi: 10.1093/molbev/msm213.

Valencia, A. and F. Pazos. 2003. Prediction of protein-protein interactions from evolutionary information. Methods of Biochemical Analysis 44: 411–426.

Waddell, P.J., H. Kishino and R. Ota. 2007. Phylogenetic methodology for detecting protein interactions. Molecular Biology and Evolution 24: 650–659. doi: 10.1093/molbev/msl193.

Wang, Z.O. and D.D. Pollock. 2007. Coevolutionary patterns in cytochrome c oxidase subunit I depend on structural and functional context. Journal of Molecular Evolution 65: 485–495. doi: 10.1007/s00239-007-9018-8.

Yeang, C.H. and D. Haussler. 2007. Detecting coevolution in and among protein domains. PLoS Computational Biology 3: e211. doi: 10.1371/journal.pcbi.0030211.

Index